Water: a way of life

Source: NASA.

Water: a way of life

Sustainable water management in a cultural context

Drs. Lida Schelwald-van der Kley

Envision-S BV, Lelystad, The Netherlands

Drs. Linda Reijerkerk MDR.

Centre for Conflictresolution, Haarlem

CRC Press
Taylor & Francis Group
Boca Raton London New York Leiden

CRC Press is an imprint of the
Taylor & Francis Group, an **informa** business

A BALKEMA BOOK

Sponsors and partners:

leven met water

PARTNERS VOOR WATER
Bundeling van krachten

UNESCO-IHE
Institute for Water Education

UNIE VAN WATERSCHAPPEN

NWB FONDS

IRC

Cover photo's:

- Courtesy of Waterschap Zuiderzeeland, NL
- Courtesy of the author

Taylor & Francis is an imprint of the Taylor & Francis Group, an informa business

Typeset by Vikatan Publishing Solutions (P) Ltd, Chennai, India.
Printed and bound in the Netherlands by the Giethoorn Media-group, Meppel.

Published by: CRC Press/Balkema
P.O. Box 447, 2300 AK Leiden, The Netherlands
e-mail: Pub.NL@taylorandfrancis.com
www.crcpress.com – www.taylorandfrancis.co.uk – www.balkema.nl

Library of Congress Cataloging-in-Publication Data
Applied for

ISBN: 978-0-415-55104-5 (Hbk)
ISBN: 978-0-203-87236-9 (eBook)

For Product Safety Concerns and Information please contact our EU representative: GPSR@taylorandfrancis.com
Taylor & Francis Verlag GmbH, Kaufingerstraße 24,
80331 München, Germany.

Contents

As President Barack Obama said in his inauguration speech highlighted the importance of water in poverty reduction:

"..let clean waters flow.....To the people of poor nations, we pledge to work alongside you to make your farms flourish and let clean waters flow; to nourish starved bodies and feed hungry minds. And to those nations like ours that enjoy relative plenty, we say we can no longer afford indifference to suffering outside our borders; nor can we consume the world's resources without regard to effect. For the world has changed, and we must change with it."

People from different countries can learn from each other. This book, therefore, highlights lessons from all continents.

Culture is a two-way street. It may act as a counteracting force or as an enabling factor for good water management. We found that many water aid projects strand because local cultural factors were not factored into the plans. We saw and learned about brand new water supply systems and water treatment installations, developed by foreign aid programs, lying idle and rusting away. Why? Because all parts of the facility that were of market value to the locals had been stolen. Or because people in power weren't pleased with the equity principle applied. Or the facility was unsuccessful just because people had to pay for water, which was in conflict with cultural and religious beliefs. Elsewhere the authorities simply didn't feel responsible and lacked the necessary skills or money to operate and maintain the installations. Corruption was found to be a killing factor for sustainable water management.

We also saw positive examples, where public participation and inclusion of local knowledge and skills in water management plans led to better health conditions and socio-economic prosperity in the region. For example, a knowledgeable woman we interviewed in South Africa put it this way: "Where people take responsibility, positive things start to happen".

The topics of culture and water management are broad in themselves, let alone the combination of the two. Sometimes it felt as if we were swimming in the middle of an ocean, the shore far away. Thanks to the critical and inspiring guidance of our Steering Group members, in particular Barber Dordregter, Sascha de Graaf, Dennis van Peppen, Bert Satijn, Henk Tiesinga, Sonja Timmer and Pieter van der Zaag, we managed to reach the shore. We also thank Deborah Sherwood and Roberta Harty for the English language editing of this book. We acknowledge Lenka Hora Adena for her supportive contributions in the final stages of the development of this book.

Wherever you live or work, we trust you will find this book an inspiration to incorporate the cultural context in your work and efforts, so as to effectively contribute to the goal of sustainable water management.

Lida and Linda

About the authors

Lida Schelwald-van der Kley

Having worked for many years as an international environmental consultant and being a member of the Board of Directors of a Dutch Water Board, Lida has become more and more intrigued with international water issues. She has noticed that despite all efforts, too many people in this world still suffer from too much, too little or too filthy water. During her more than 20 years of consultancy work for both private companies and government organizations, Lida has experienced that communication and cooperation are important for the success of any project. Taking into account the cultural context is likewise important.

She wondered if an exploration of the relationship between water and culture world-wide could help to find the keys to successful and sustainable water management. Lida decided to start an explorative journey across the continents and to lay down her findings in a book. At this point she met an "old" colleague, Linda, who brought along a wealth of world-wide cultural expertise, and they decided to join forces.

Lida has written several international publications on successful communication between industrial companies and their stakeholders. One widely used, as it provides practical guidelines, is "Communication on contaminated land management". Since 1994 Lida has been the Managing Director of Envision-S.

Lida was born in Vancouver, Canada and has lived and worked most of her life in the Netherlands.

She is married to René and they have two children, Esmay and Rayner.

Linda Reijerkerk

Since 1985 Linda Reijerkerk has been working as a cultural anthropologist and an international consultant in water management with a clear focus on processes and institutional development and management. She has worked in Western and Eastern Europe, Africa, Asia and South America for institutions including the World Bank, Islamic Development Bank, EBRD, African Development Bank and the Dutch Ministry of Foreign Affairs. On all of these continents her work has always focused on project management, institutional development, training and communication/education. Linda speaks Spanish, Portuguese, English, German, French and Dutch.

As Linda saw an increase in environmental conflicts, she specialized as a mediator in 1997, when mediation was still a novelty in the Netherlands. She is now Director of the Centre for Conflict Resolution in the Netherlands, one of the top three Dutch companies specialized in trainings for mediation and negotiation. She is furthermore a professional coach and mediator for business and workplace-related disputes, as well as for public disputes (environmental, multi-party mediation) with a special interest in water conflict resolution.

She is co-author of, among other publications, the *Praktijkgids Mediation* (Practical Guide to Mediation, 2005), *Diversity Management* (2005), and *Intercultural Conflict Management and Mediation* (2009). She has been a board member of the Dutch Mediators Association (NMV) and is presently President of the European Mediators' Network Initiative, EMNI.

Linda is married to Jon Schilder and they have two children, Florian and Amaya.

1 – INTRODUCTION TO WATER & CULTURE

Chapter 1

Introduction to water & culture

> Our history is tied to these waters. Our continued reliance on fishing, trapping and hunting and our desire to do so is dependent on these waters. Our future is based on these waters... Any threat to such waters poses a direct threat to our survival.
>
> Grand Chief B.G. Cheechoo, chief of the Nishnawbe-Aski Nation in Canada, explaining the relationship between water and his culture.

Have you ever wondered why water management projects often fail? Are you concerned about the fact that as a consequence too many people in this world still suffer from too much, too little or too filthy water? Are you looking for sustainable solutions for water resource management that really work? If so, we hope you find this book about water and culture a source of inspiration.

This book, which is based on extensive research, is intended to form a cultural bridge towards new sustainable water management practices. *Water: a way of life* takes you on a water journey through time and across the world's continents. Along the way it explains the past and present ways in which different cultures around the world, both traditional and modern, have viewed and managed water in response to the environment they inhabit. A better understanding of these cultural water beliefs and practices may lead to new concepts for future sustainable water management – from flood management to water supply, sanitation and irrigation management.

The book may prove useful to water professionals exporting knowledge and technologies to foreign countries. The challenge is to come up with sustainable solutions for water management by taking into account local, cultural factors. The book is also meant to encourage world leaders, politicians and decision makers responsible for water management to really make a change for the benefit of the people they represent. The book provides clear examples and answers on how to improve sustainable water management by taking into account cultural aspects and the way in which these influence individuals *and* institutional decision-making processes.

1.1 THE IMPORTANCE OF THE CULTURAL DIMENSION

Why is the cultural dimension so important in sustainable water management?

Back in 2003, at the third World Water Forum in Kyoto, UNESCO issued a Statement[1] in which they clearly indicated the importance of the relationship between

Figure 1.1 Living with water.

water and culture. It said, among other things, that:

- Water is a vital resource, having economic, ecological, societal and spiritual functions. Consequently, its management greatly determines sustainability.
- Due to its fundamental role in the life of societies, water has a strong cultural dimension. Without understanding and considering the cultural aspects of our water problems no sustainable solution can be found.
- Relations between peoples and their environments are embedded in culture. The intimate relationship between water and peoples should be explicitly taken into account in all decision-making processes.
- The ways in which water is conceived and valued, understood and managed, used or abused, worshipped or desecrated, are influenced by the cultures of which we are a part.
- As the frequent failure of "imported" solutions has proven, water resources management will fail if it lacks the full consideration of these cultural implications.
- Cultural diversity, developed over millennia by human societies, contains a trove of sustainable practices and innovative approaches.

As such, managing water is as much cultural as it is technical. For these and other reasons, the cultural dimension of water and its management deserves further exploration.[2]

As indicated in various papers of the World Bank, 'Culture is not an optional special interest, nor a sector in administrative terms. It is rather a critical filter through which the content, design and expected impact of the Bank's development activities should be examined for improved effectiveness and outcome.'[3]

1.2 STRUCTURE OF THE BOOK

The chapters in this book touch upon a number of cultural aspects and topics relevant to achieving the goal of sustainable water management. Each chapter provides a mix of noteworthy examples as well as historic and recent facts about water management in relation to culture. They are further illustrated by case studies and views from visionary people telling their sparkling stories about water and culture.

The five main chapters pertain to water as:

1.2.1 A source of life

Water and culture are strongly interlinked. Water is a vital *source of life*, and culture greatly determines a person's "way of life." This includes the way people manage their water resources, having adapted themselves to the environment in which they live. For centuries sustainable water management has been daily practice in many cultures, resulting in a delicate balance between water resources and human society. Water has played a more or less prominent role in cultures, depending on the environmental conditions people had to face. Indigenous cultures are still renowned for their ingenious and sustainable water practices. Modern practices have often disturbed and overruled these traditional practices, with undesired consequences. Examples can be found in many former colonies, where social structures were disrupted by the foreign rulers. Despite the high quality of many of the pre-existing water management systems, the new rulers considered them to be primitive and "backward." They used their supposed superiority to urge natives to exploit the valuable water resources. Water lost its mystery and came to be seen as a mere commodity. However, in today's societies we often see a revival of old traditions and a more natural and sustainable use of water.

1.2.2 A source of inspiration

In most cultures water is also *a source of inspiration* and has been for many centuries. People have adopted deeply rooted spiritual and religious values and beliefs that bind them and support them in living the way they do. These play an important role in water management. Yet despite the water wisdom we find in many societies, wise water management is still a bridge too far – or isn't it? Lately, religious and spiritual leaders have been rethinking their values and their roles in safeguarding creation and the earth's resources.

1.2.3 A source of power

For a long time water was viewed as *a source of power* that should be controlled. The Egyptians and Romans set the example, being masters in irrigation and the creation of major water works that controlled water's ways. In more recent time periods large-scale dams for water storage and hydropower became common practice, thereby often displacing entire communities and ecosystems. Furthermore, freely meandering rivers were engineered and tamed to become artificial "rivers" (canals). Many countries are currently trying to reverse this latter practice by allowing rivers to once again meander freely.

Throughout the ages water has also be used to exert political power and gain control over people and over entire regions.

1.2.4 A source of conflict or of cooperation?

Both within and between states, access to water and control over its allocation can become *a source of conflict*. This naturally becomes most evident in regions where water is scarce and the demands from different users are high. A well-known example is the Tigris-Euphrates river basin. As a result of the construction of dams, hydro-electric power plants and large-scale irrigation works upstream in Turkey, the flow of water to the downstream countries of Iraq and Syria is reduced considerably. This obviously causes tensions between these countries, as water in the region is already short in supply. If we look at it from the positive side, shared water resources between countries that share the same river basins can also provide opportunities to promote international cooperation – as opposed to interstate conflict. As such, water can be *a source of cooperation.*

1.2.5 A source of sustainability

Cultural traditions cannot be ignored when trying to solve the global water issues and challenges facing us today. These are driven by either a scarcity or an abundance of water and are exacerbated by climate change.

The ongoing challenge, from the local to the global level, is to take appropriate measures to be able to foresee the need for water at the right time and place and of the right quality for the benefit of all. Population growth and the resulting increase in water demand and pollution, as well as the expected sea-level rise due to the resulting climate change, call for adequate and timely action. World-wide problems of droughts and flooding ask for international cooperation. In fact, our future depends on it. Having a "common challenge" to mitigate and adapt to climate change may help, as throughout the ages people have realized that cooperation is essential for survival. Furthermore, a better understanding of the way various cultures in different parts of the world perceive and manage water may lead to new concepts for sustainable water management. Water needs to again become *a source of sustainability.*

1.3 THE BROADER CONTEXT OF SUSTAINABLE WATER MANAGEMENT

'Knowing the context is knowing the kind of questions to ask.'

1.3.1 Sustainable water management

In order for water management to be sustainable, a balance needs to be found between social, environmental and economic factors. This notion of sustainable development, often referred to as the *Triple P* concept, has been embraced by many around the globe. It aims at development that meets the needs of the present without compromising the ability of future generations to meet their own needs.[114]

How do culture and water management fit into this picture? To start with the latter, the sustainability concept is seen as protecting and managing water as a valuable resource of the earth for the use of present and future generations, while at the same time balancing the different – and sometimes conflicting – interests and needs. As stated in the introduction of this book, water has economic as well as ecological and social functions. The challenge is to find a sustainable balance between these aspects.

Culture has everything to do with people, but also with the interaction people have with their environment, including the way they view and use their water resources.

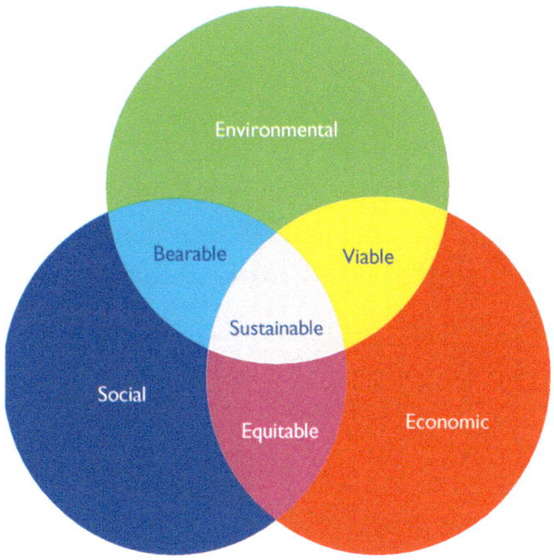

Figure 1.2 Sustainability is about finding a balance between social, environmental and financial needs and realities.

> **Finding the right balance...**
> There are three main pillars of sustainable development, often represented in the well-known and broadly used Triple P (**P**eople, **P**lanet, **P**rofit) concept. These pillars are also referred to as social, ecological and economics conditions.

Ideally, water management practices should be socially and environmentally *bearable,* thereby not exceeding the carrying capacity of the environment. Next there are the socio-economic relationships between people and profit-making. These should also be taken into account in decision-making processes about water management, with an aim for *equitable* solutions. Water is the fuel for our economy. Water can be a limiting factor in socio-economic development but also an opportunity for people to obtain a better livelihood. Trying to find a fair balance entails the risk of trade-offs between the three pillars of sustainable development. As such, ecologies and economies often seem to be in conflict with each other, also when it comes to water management. However, it is now widely acknowledged that ensuring a *viable* balance between the two stands the highest chance of success in the long term and will lead to sustainable wealth and well-being.

1.3.2 The broader context

In order to achieve the Millennium goal of safe drinking water and sanitation for all, but also to guarantee safety against too much water, sustainable water management is a prerequisite.

In general, sustainable water management implies the need for:

- sustainable exploitation of water resources;
- effective quantitative and qualitative management of water;

- efficient distribution and use of water for domestic, agricultural, commercial and other purposes;
- controlled discharge and treatment of waste water (sewage, etc.); and
- adequate flood protection.

Any new plans for sustainable water management should fit into the broader context, which is often culturally determined. Water management has its own domains and values, each of these having its own historical dimensions. Apart from the technical aspects, it deals with:

- Ecological aspects;
- Political aspects;
- Economic aspects;
- Legal aspects;
- Institutional aspects; and
- Social aspects and cultural aspects.

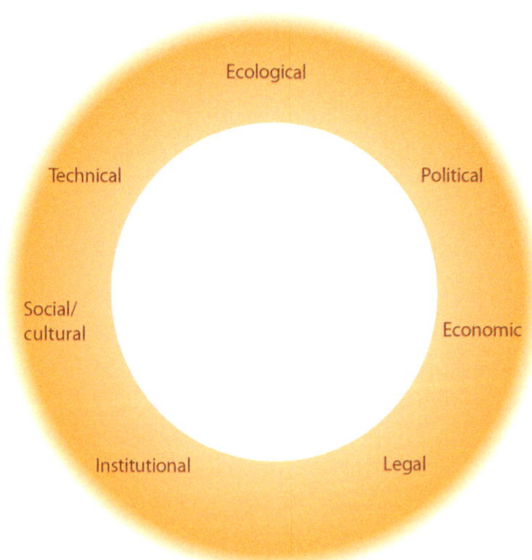

Figure 1.3 The broader context of sustainable water management.

Each one of these domains influences the other ones. They are in systemic relationship to each other, this applying most strongly to the ecological and socio-economic aspects. For example, when cattle holders move to a new area, this has an impact on the ecology, especially in cases of overgrazing. Another example is how new "national" borders in colonial Africa have led to increased water use in some areas and reduced use in others.[4]

Furthermore, the concomitant values belonging to a particular domain may conflict with the values belonging to another domain. For instance, the equal distribution of water (in this case a legal or social norm) may go against the social norms of a traditional hierarchical society. An example of this situation is a water irrigation scheme in the Marib area, where a new, more equal distribution of water led to social disruption and conflict.

Performing an assessment of potential cultural impact prior to the installation of a new system or infrastructure can prevent such undesired consequences from happening. It can be a useful tool to ensure the sustainability of a culture and the cultural heritage of a society.

Context of sustainable water management and issues to consider

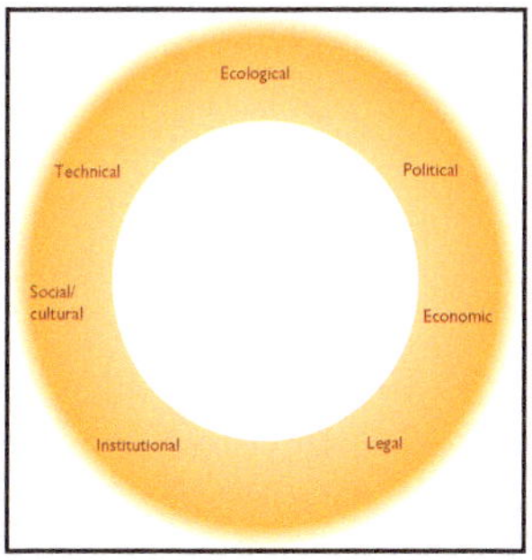

Technical context

- What technical solutions are used and accepted in the culture in question, and which have proven to work well under these circumstances?
- Is it possible to integrate these solutions in future water management concepts?

Ecological context

- Is there an existing sustainable relationship with water resources that can be built upon?
- What is the impact of your plans on the ecosystem?

Political context

- Is the political regime centrally led or organized in a decentralised manner? Is it stable or unstable, supportive or not?
- How does the political culture foster or negatively influence the sustainability of water systems?

Economic context

- What kind of financial resources or access to external funding is available toward achieving sustainable water management?
- How will a water facility (dam; driving watersupply) effect the economy?

Legal context

- What kind of legislative framework is in place, and is it adequate for improving water management?
- How do traditional water rights influence water sustainability?

Institutional context

- How is water resource management organized?
- Can you involve local organizations in your plans?

Social context

- What social structures should be considered when designing a water system or facility, dam, etc.?
- What is the socio-economic impact of your plans?

1.4 CULTURE

1.4.1 The cultural context

Decisions regarding water in any country are often made on the basis of the dominant cultural values in that society. The same applies to the acceptance of solutions. This is important to realise when exporting 'technological solutions' from one country to another. One has to take into account the context of values in which these solutions were developed and the context of values in which they will land. The saying "*What works in one place doesn't necessarily work in another*" applies.

1.4.2 What exactly is culture?

There are many definitions of culture. In cultural anthropology, culture is often referred to as 'the system of shared beliefs, values, behaviour and symbols that the members of society use to cope with their world and with one another, and that are transmitted from generation to generation through learning'.[5] In short, it is the way we think, feel and act within the boundaries of what the social context determines as acceptable.

Within a country several subcultures or ethnic groups may exist. For instance, in Indonesia hundreds of ethnic groups co-exist, with almost 300 languages being spoken. Also, cultures often cross several country boundaries. The Kurds, as an ethnic group, are indigenous to a region that includes adjacent parts of Iran, Iraq, Syria and Turkey. At the same time there can be large differences in individual behaviour amongst people living in the same cultural society. To make matters even more complicated, a culture is not a static phenomenon. Cultures change – some more rapidly than others – and societies change through modernization and globalization.

1.4.3 Understanding culture

Understanding and considering the cultural aspects of water problems is necessary in finding a sustainable solution. This was painfully depicted in a yet unpublished federal history report[6] of the American-led reconstruction of Iraq, aimed at, among other things, the provision of water to local households. Ignorance of the basic elements of Iraqi society and infrastructure, which are controlled by neighbourhood politicians and tribal chiefs, was a main cause of failure. Any new plans, whether they be reconstruction, relief or commercial plans, should fit into the cultural context of a society. These include cultural and religious beliefs, values and practices.

In order to understand the values and beliefs of a culture, we must get to the heart or core of it, which is by no means simple. Cultural values and beliefs are usually deeply rooted, as symbolized below by the Nautilus shell.

The way in which cultures manifest themselves is made tangible through:

1 Values and beliefs;
2 Traditions, rituals and practices; and
3 Symbols and artefacts.

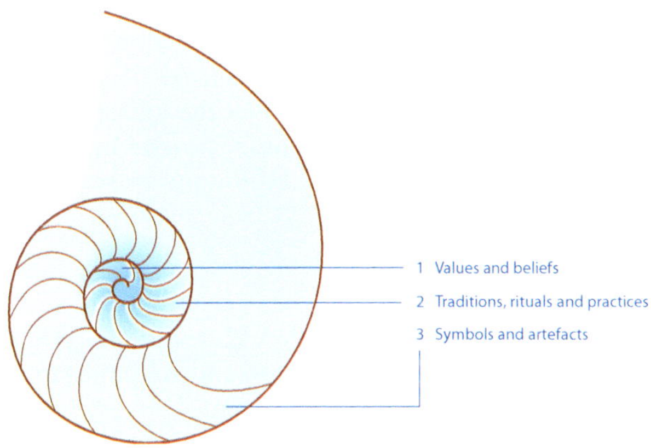

Figure 1.4 The Nautilus shell symbols different layers of culture.

> The Nautilus lives in a watery, marine environment and is in constant interaction with the outside environment. The growth of its spiraled shell starts from the centre. New and bigger chambers are formed with a lunar month periodicity. The shell finishes with a large funnel where the animal itself lives. Through a "life-cord" it stays in connection with its inner core.

1. Values and beliefs

Values, norms and beliefs form the core of any culture and are usually not made explicit or visible. They have to do with feelings about what's good or bad, dirty or clean and abnormal versus normal.

EXAMPLES FOR WATER MANAGEMENT

A river we consider filthy and unhealthy may not be regarded as such by locals, living adjacent to it, who use the water for many different purposes. In such a case it is important to find out why a certain source is preferred.

As another example, according to cultural and religious beliefs, many cultures adhere to the principle that water use should be free of charge. This widespread notion, which is encountered in Islamic countries and many traditional cultures, makes it difficult to convince people to pay for water supply services delivered. To deal with this, a distinction is often made between the water itself and the services delivered for transport and treatment. Different tariff systems may also be applied to water for religious purposes and for domestic or commercial uses.

2. Traditions, rituals and practices

Each culture has its unique set of traditions, rituals and practices. Water is often a key element of these. Examples are the rain dances in Africa and amongst first nation peoples in the United States that are performed to ensure water for people, cattle and crops; or the numerous water rituals in Japan, believed to provide relief or assurance for almost any event in one's life. Throughout the ages people have also developed and used traditional methods for water distribution, conservation, protection, etc., that have allowed them to survive in their particular climate and environment. Therefore, cultural diversity, developed over the millennia by human societies, in itself constitutes a treasure of sustainable practices and innovative approaches to water management. These can sometimes be built upon or revitalized for future sustainable water management.

EXAMPLES FOR WATER MANAGEMENT

A good example of the revival of sustainable old traditional practices is the Karez system in arid parts of countries such as Pakistan and Afghanistan. These are ancient underground water and irrigation channels, designed to overcome the problem of evaporation losses. They are fully owned, operated and maintained by the local community and therefore fit into a collective society, in which people are used to collectively sharing responsibilities and the workload. Nowadays many have been restored to their original state. This will be elaborated on in the next chapter.

An example from the West is the Dutch "polder" system. The local communities worked together to establish levees and dikes in order to reclaim land from the North Sea on which to live and work.[7]

3. Symbols and artefacts

Symbols and artefacts, like words, gestures or pictures, are expressions of culture. As such, they are more visible on the surface. Words and expressions used for water also tell us a lot about the importance attached to water in a particular culture. For example, a most fascinating aspect of the native language of the Inuit in the Arctic region is the language's numerous words for snow. The Inuit make a distinction between falling snow (*annui*), snow that sticks to the branches of trees (*qali*) and snow with a surface of very smooth and fine particles (*saluma roaq*). Many cultures also have heroes. These are persons, dead or alive, revered by a society. In the "water world" Hansje Brinker, who supposedly stuck his finger in a dike to prevent a Dutch town from flooding, still symbolizes the Dutch struggle against water. Most first nation societies in the US and South America have heroes, proverbs and stories about water.

Expressions of culture and using stories to convey messages can be of value to water professionals, especially professionals in the field of water and sanitation education.

EXAMPLES FOR WATER MANAGEMENT

In Sierra Leone the concept of the Juju man was used to convey messages regarding hygiene and water management. The Juju man is a kind of witch doctor, common in West African countries, originating from the Yoruba in Nigeria. As an important and respected figure in local communities, he is an outstanding person to convey a message about water management and hygiene.

Dams and power plants are pre-eminently symbols of power, often used by powerful rulers to express their supremacy and the power of men over nature.

The next chapters further explore the relationship between water and culture.

2 – WATER: A SOURCE OF LIFE

Chapter 2

Water: a source of life

A great mother named Mekong

She touches so many lives bringing forth life. China, Vietnam and Cambodia respectfully address her as The Great Mother. And even though we think we can stand on our own, we keep going back to her for nourishment.

The great Mekong River provides the food for millions of families, caressing the land so that crops and fruit trees grow. We find ways to purify her waters to quench our thirst. She takes us places, transfers goods and attracts people from around the world to marvel at her beauty.

Mother Mekong has another side to her that people fear. At times she displays her anger and might, pouring her waters over huge tracks of lands. Some people refuse to give way. They challenge her strength by surrounding their villages with walls and fortresses. Most of the time this is to no avail and Mother Mekong whips them where they least expect it. They suffer for their stubbornness. Other children know it is wise not to stay too close, building their homes beyond her stinging lashes. The wisest of her children let her anger pass through underneath their stilt houses, knowing that her mood will pass.

But mighty as she is, the Mekong is showing her own vulnerability. We have filled her waters with trash of all kinds, thinking she could cope with it all, but day by day we are poisoning her. As she chokes her life force ebbs away. We rob her of her role as a life giver.

Mekong is a complex being. But, in reality, it does not take a genius to understand her, if we take the time to watch her moods and her health. As children we have taken from this mother. Now, as we grow we must learn a give and take relationship that will protect both mother and child.

For centuries the Mekong River looked after herself, as she looked after us. Today, we must learn to look after her in return. We are all sons and daughters of the water.

Charles Bautista, the Philippines[8]

2.1 INTRODUCTION

This chapter starts with a description of the uniqueness and presence of water and its importance for human survival. It then explores the ways in which cultures all over the world have adapted themselves to living with water under different circumstances with the sometimes limited means they have at their disposal. Nowadays humankind is facing new challenges in a changing world, having major consequences for water management. A question to be answered is: Can we incorporate the old "proven" adaptation mechanisms into new sustainable water management strategies?

Source: NASA.

The distribution of water on earth: omnipresent, yet scarce

There is an abundance of water on our planet. Seen from the sky, the earth looks indeed like a "blue planet". Yet 97.5% of the earth's water is salt water and can be found in the oceans and seas. Only a mere 2.5% is fresh water, and most of this is "trapped" in glacial ice and the polar caps. Less than 1% of fresh water exists in lakes, streams, rivers and shallow underground reservoirs that can be used as drinking water, industrial water or as irrigation water for food production. Moreover, fresh water supply is not evenly distributed on earth. The climate and amount of rainfall largely determine the availability of water. Only a limited number of countries are lucky enough to have enough fresh water resources to meet the demands of their people. Climate change will further diminish the availability of fresh water.

2.2 WATER: THE FIRST AND FOREMOST (RE)SOURCE

2.2.1 A recurring resource

Life on our planet began through the presence of water, and life can continue because of it. Water is universally recognized as a vital source for humans and ecosystems, although its uniqueness is sometimes overlooked.

Just imagine that the water you drank a few minutes ago may have been around for millions of years. Its water molecules have evaporated from the sea and fallen from the sky as rain on the earth or snow on the mountains. They then melted to form a whirling stream and finally calmed down to cross village communities as a meandering river through a green valley. Yet the story doesn't end here. The water may have flown back to the sea or could have been used by farmers to water their crops. It became trapped in underground reservoirs and was finally pumped up and used by you as drinking water to re-enter the water cycle yet again. In fact, it may be the same water that your pre-ancestors drank hundreds of thousands of years ago. If water could tell what it had witnessed in the meantime... Time will tell.

For the time being we will have to rely on oral traditions and written knowledge to tell us more about cultural practices that can help us today to make sound and sustainable decisions for future water management.

2.2.2 A cradle of civilization

From the very beginning of our human history, water has been recognized as a precondition for human settlement. Our pre-historic ancestors, mostly "hunters, fishers and gatherers" already understood the advantage of settling themselves close to bodies of water. They used the water for cooking, cleaning and food production and – not to forget – for transportation. The abundant availability of water throughout the world in fertile river

Source: NASA.

The distribution of water on earth: omnipresent, yet scarce

There is an abundance of water on our planet. Seen from the sky, the earth looks indeed like a "blue planet". Yet 97.5% of the earth's water is salt water and can be found in the oceans and seas. Only a mere 2.5% is fresh water, and most of this is "trapped" in glacial ice and the polar caps. Less than 1% of fresh water exists in lakes, streams, rivers and shallow underground reservoirs that can be used as drinking water, industrial water or as irrigation water for food production. Moreover, fresh water supply is not evenly distributed on earth. The climate and amount of rainfall largely determine the availability of water. Only a limited number of countries are lucky enough to have enough fresh water resources to meet the demands of their people. Climate change will further diminish the availability of fresh water.

2.2 WATER: THE FIRST AND FOREMOST (RE)SOURCE

2.2.1 A recurring resource

Life on our planet began through the presence of water, and life can continue because of it. Water is universally recognized as a vital source for humans and ecosystems, although its uniqueness is sometimes overlooked.

Just imagine that the water you drank a few minutes ago may have been around for millions of years. Its water molecules have evaporated from the sea and fallen from the sky as rain on the earth or snow on the mountains. They then melted to form a whirling stream and finally calmed down to cross village communities as a meandering river through a green valley. Yet the story doesn't end here. The water may have flown back to the sea or could have been used by farmers to water their crops. It became trapped in underground reservoirs and was finally pumped up and used by you as drinking water to re-enter the water cycle yet again. In fact, it may be the same water that your pre-ancestors drank hundreds of thousands of years ago. If water could tell what it had witnessed in the meantime... Time will tell.

For the time being we will have to rely on oral traditions and written knowledge to tell us more about cultural practices that can help us today to make sound and sustainable decisions for future water management.

2.2.2 A cradle of civilization

From the very beginning of our human history, water has been recognized as a precondition for human settlement. Our pre-historic ancestors, mostly "hunters, fishers and gatherers" already understood the advantage of settling themselves close to bodies of water. They used the water for cooking, cleaning and food production and – not to forget – for transportation. The abundant availability of water throughout the world in fertile river

Chapter 2

Water: a source of life

A great mother named Mekong

She touches so many lives bringing forth life. China, Vietnam and Cambodia respectfully address her as The Great Mother. And even though we think we can stand on our own, we keep going back to her for nourishment.

The great Mekong River provides the food for millions of families, caressing the land so that crops and fruit trees grow. We find ways to purify her waters to quench our thirst. She takes us places, transfers goods and attracts people from around the world to marvel at her beauty.

Mother Mekong has another side to her that people fear. At times she displays her anger and might, pouring her waters over huge tracks of lands. Some people refuse to give way. They challenge her strength by surrounding their villages with walls and fortresses. Most of the time this is to no avail and Mother Mekong whips them where they least expect it. They suffer for their stubbornness. Other children know it is wise not to stay too close, building their homes beyond her stinging lashes. The wisest of her children let her anger pass through underneath their stilt houses, knowing that her mood will pass.

But mighty as she is, the Mekong is showing her own vulnerability. We have filled her waters with trash of all kinds, thinking she could cope with it all, but day by day we are poisoning her. As she chokes her life force ebbs away. We rob her of her role as a life giver.

Mekong is a complex being. But, in reality, it does not take a genius to understand her, if we take the time to watch her moods and her health. As children we have taken from this mother. Now, as we grow we must learn a give and take relationship that will protect both mother and child.

For centuries the Mekong River looked after herself, as she looked after us. Today, we must learn to look after her in return. We are all sons and daughters of the water.

Charles Bautista, the Philippines[8]

2.1 INTRODUCTION

This chapter starts with a description of the uniqueness and presence of water and its importance for human survival. It then explores the ways in which cultures all over the world have adapted themselves to living with water under different circumstances with the sometimes limited means they have at their disposal. Nowadays humankind is facing new challenges in a changing world, having major consequences for water management. A question to be answered is: Can we incorporate the old "proven" adaptation mechanisms into new sustainable water management strategies?

The structure of water and its unique properties

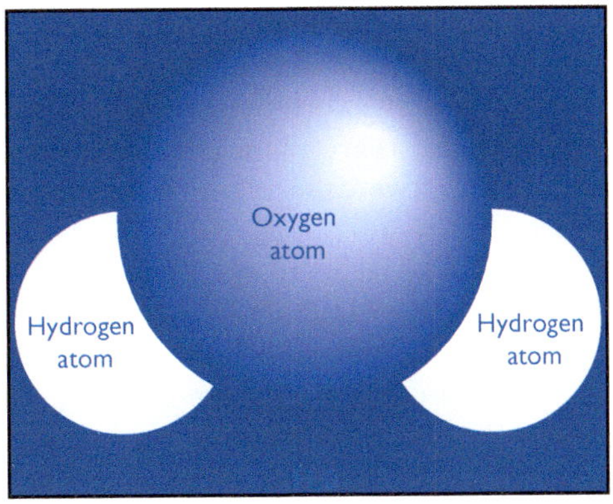

Although the structure of a water molecule (H_2O), with two hydrogen atoms and one oxygen atom, is in fact quite simple, its properties are mostly unique. To mention just a few: Water is able to absorb and release heat, thereby keeping Earth's temperature shifts within limits that permit life. The same applies to our body temperature, as our bodies consist for roughly two-thirds of water. Plant and animal life have also benefited from water's unique properties. Thanks to its cohesion and adhesion properties, water can be transported against gravity, and thus it became the prime transportation mechanism in the ecological society.

locations created the right conditions for nomadic people to form sedentary, agrarian communities, therewith becoming "cradles of civilization". Amongst historians there is debate about the exact location of the first civilization, as it is claimed by several regions. The controversy is complicated by the difficulty of defining when exactly a culture becomes a civilization, having to do with the development of writing, forming complex social systems and cities. Yet there is no discussion about the fact that it was rivers that created the conditions for great civilizations to emerge. The Nile fulfilled this role for the Egyptians, and the Mesopotamians thanked their highly civilized society to the Euphrates and Tigris, drawing water from these mighty rivers to irrigate their land. Yet scarcity of water during extended periods of droughts along with mismanagement brought down these same powerful societies.[9] Other rivers associated with the cradles of civilization are the Indus in South Asia and the Huang-He-Yangtze River in China. But even today large metropolises like Rotterdam, London, New York City and Shanghai owe their successful expansion to their easy accessibility via water and the resulting trade business.

2.2.3 The primordial place of water

Water's central place is also recognized in philosophy, in most world religions and in the cosmic perception of many cultures. Greek philosophy begins with the notion that *water* is the primal origin of all things. Thales of Miletus most famous belief from about 580 BC was his cosmological doctrine, which held that the world originated from water. In the same way, water and its primordial place is venerated in many religions, as will be further described in the next chapter, 'Water: a source of inspiration'. As for the origin of life, The Old Testament of the Bible states in the book of Genesis that 'Water was created on the first day' and in the book of Exodus that 'Water brings life'. In the same respect the Koran mentions that 'All life originated from water' and that man himself is created of water, as are all the animals on earth. Indeed, our bodies are made largely of water, and as we cannot live without water it is a primordial human right by our very nature.

In the Andes the local people call it the 'blood of the earth', the source of life from which other life grows. On the same South American continent subsequent generations of conquerors, from Inca emperors to invading Spanish Conquistadores, gave water a central place in their cosmic view to claim their supremacy.[10] The gods that the Incas worshipped lived in the same mountains as from where the water originated, therewith positioning themselves in the centre of the universe. In addition, the Inca rulers elevated themselves by excluding the subordinated classes, mostly common Indian people, from the benefits of society, including access to vital water resources. Today, across the Andes and in many other regions throughout the world, the inclusion and participation of excluded, often indigenous communities is gaining more support as being a wise political strategy. As long as this strategy takes into account the diversity of cultures, values, views and different water management practices and incorporates them respectfully, it can lead to something productive. This also applies to participatory water management, where local people are actively involved in decision-making around local and regional water management issues. Community management of water, where people bear responsibility for local water resource management, greatly enhances the acceptance and sustainability of solutions.

Case study

The right to be different

In the Andes mountains water is there for the use of the communities. It's not up to individuals, corporations or the state to buy or sell it, to take or give it away. The right to use the water can only be earned by taking responsibility for a fair and sustainable distribution. By contributing to the design, the construction and organization of the irrigation system, the users create water rights.

In the Licto area, near Riobamba, in Ecuador, the indigenous population fought for its water. Indian peasants headed the design, construction and organization of the irrigation system. After more than 20 years the water finally reached the community.

Ancient and modern conquerors of the Andean highlands denied the indigenous people access to springs and rivers. *Water became a source of conflict*. And usually the Indians got a raw deal. For the indigenous and peasant water users there is more at stake than just getting access to water. The authority to manage the system according to their own set of rules is what they demand. Invariably, local water management has been neglected and suppressed by national policies and legislation. Thus in the end the indigenous peasant fights not only for water but for recognition of their water rules, authority and culture as well.

In the 1990s the town of Licto, dominated as it was by the big white and mestizo farmers, refused to work together with the Indians. Inés Chapi, an Indian woman, who came a long way from being oppressed and discriminated against to becoming a more respected irrigation organizer and peasant leader, supported the formation of an irrigation committee in town and was elected as a board member. They forged a strategic alliance with the indigenous peasant organization. Gradually the mestizo resistance waned away.

But the authority of the indigenous people to run the system is challenged time and again. Government wants to take over the control of the system, now that it's running smoothly. And conflicting interests are bound to lead to clashes. The battle is not over yet.

Based on: 'The rules of the game and the Game of the Rule', by Prof. Rutgerd Boelens, Wageningen University, NL, 2005.

2.3 THE SECRET TO SURVIVAL...

'*Where there is water, there is life*' is a well-known saying. Across the world people have adapted to living with water in many ways in response to the environment they inhabit. This section describes a cross-section of the most distinct environments on earth and the way cultures live and deal with water in these areas.

2.3.1 Living with water in the mountains

Over 1 billion people in the world live in mountainous regions and have adapted their distinct lifestyles, remote from the rest of the world, to the mostly harsh and cold climate. Their cultures are often quite characteristic, as can be seen in the Andes or in the Himalayas. In mountain communities a strong sense of the need for interdependence to survive the harsh climate has led to distinct cultural traits, such as a collectivistic society and living in a harmonious relationship with one another.

From a water point of view, mountains are globally important as the source of most of the earth's fresh water. Mount Kenya, as an example, is the prime source of water for more than 2 million people in Africa.

However important, this seems insignificant compared to the majestic Himalayas, often referred to as 'the roof of the world' and also known as 'the water tower of the world'. The Himalayas are the main water source of 9 giant river systems of Asia: The Indus, Ganges, Brahmaputra, Irrawady, Salween, Mekong, Yangtze, Yellow, and Tarim.[11] As such, they are the water lifeline for over 500 million inhabitants of the region. Yet despite the proximity to the water source, the living standards of the people living at high altitudes in the mountains are generally poor, and obtaining safe and reliable water can be a daily challenge.

Living, for instance, high up in the cold mountain deserts of Ladakh, it can be a daily challenge to obtain water. The streams are frozen year-round, and when they finally melt and start flowing for a short while they bring along loads of sediments from the glaciers. An option is to melt snow or ice, but this requires fuel, which is hard to obtain in these remote areas, and hauling water from lower elevations is too labour-intensive. Modern solutions to alleviate the burdens of mountain people in need of reliable water concentrate on drilling boreholes at high altitude to extract water from subsurface groundwater reservoirs. This has proven to be a successful

Figure 2.1 Living in Ladakh.

strategy so far and has brought relief to the people living in these mountainous regions.

An innovative approach of water harvesting in Ladakh involves bringing glaciers closer to the villages and farm lands. This is done by channelling water from a local stream to an area in the shady side of the mountain, so that it freezes in winter. It thus creates an artificial glacier close to the village that melts already in spring, at the start of the sowing season, when the farmers need water for irrigation.

In less elevated (and less cold) mountain areas like the Rocky Mountains in Canada and the Unites States or the Alps in Europe, people living in mountain villages tend to mainly benefit from being close to the water source. They have a reliable and high quality water source at hand and at the same time benefit from using the water gradient as a cheap and relatively clean hydropower source. More of the use of water as a source of power is described in Chapter 4.

2.3.2 Roaming the rivers

Rivers have always been ideal places for human settlement. Rivers provide riparian communities with water for each and every purpose, no matter in what part of the world they flow. Take for instance the Surinam River in the South American country of Surinam. In the upstream regions of this river many so-called "Marron" communities inhabit the river banks. They are the descendants of former West African slaves who fled the inhumane working conditions at the plantations and used the river to escape. As almost eighty percent of Surinam consists of inaccessible primary rainforest, settling themselves close to the rivers was their only chance of survival. Local Indians, living further inland and deeper in the forest near the creeks, taught the Marrons how to survive in the jungle and use the plants, trees, animals and water for their daily needs, thereby living in close harmony with nature. Most of these Marron

Figure 2.2 Water has an important social function in Surinam.

Figure 2.3 Likewise in Western Africa.

communities are still relatively untouched by modern influences, and they live a sustainable and authentic way of life. Their culture and use of water still bears a strong resemblance to West African cultures. The people use the "living" water from the river and creeks for drinking, cooking, bathing and washing.

Many hours of the day you can find the women and girls at the river banks fetching water, bathing themselves or cleaning clothes and scrubbing pots and pans with a mixture of sand and water, while chatting and socializing with one another. Water obviously has a strong social function for them.

Further downstream and coming closer to the city of Paramaribo, the Marron communities have been influenced by the modern world. In some of these villages drinking water supply systems have been installed, sometimes in vain. In one instance a modern system was installed in 2005, but due to an unforeseen major flood a year later, the system became flooded with sediment-filled water; it rusted and now lies idle. As water used to be free and because the people do not have much money, no one from the community is willing to or feels responsible to invest in a new system. In another village a "tap" with a ceramic sink was installed in the middle of the village. The women soon started using it for every purpose, such as doing their laundry, but also cleaning the chicken they had just killed for dinner. The contaminated water reappeared above the surface at a short distance from the tap point, flowing through open gullies amongst the wooden and palm tree houses. This attracted mosquitoes and other vermin, thus becoming a source of disease. Realizing this, the women adapted some of their practices. New forms of waste form another threat to the water quality and the precious rainforest ecosystem. Cola in plastic bottles, diesel containers for the generators and other modern joys have recently found their way to most of the riparian Marron communities. People are used to throwing away any waste products, such as nut shells, around them. This is not a problem as long as it is biodegradable organic matter. There is insufficient awareness, however, of the fact that plastics and chemicals can harm the environment. Bauxite mining and gold mining with cyanide and mercury along the Surinam River form a particularly serious threat to the water quality and ecosystems of the rivers and creeks and the people living along the river banks. Humans can be exposed through direct contact, intake of water and fish consumption. Economic considerations tend to override ecological and health concerns, something which may backfire in the end.

The rest of the world is not doing much of a better job. The water quality of most major rivers in the world is under threat, especially in booming industrialized countries like India and China. Even in modern Europe, where human societies and industrial activity have long prospered along the banks of the rivers, there are still villages, cities and industrial plants that do not have proper waste water treatment facilities, discharging their pollutants directly into the water. Water legislation introduced in many North-Western European countries in the 1970s reversed this trend. Salmon even returned to the River Rhine, indicating the amelioration of the water quality. It is notable that in the past many houses used to turn their backs to the filthy rivers, whereas most houses now face the water. Water is once again appreciated as a valuable asset. The new European Water Framework Directive, which was implemented in 2000, prescribes river basin management for all European river systems and stresses the need for cooperation on a river basin level. This will further help to improve the water quality of the rivers in Europe.

2.3.3 Dealing with water in delta areas

Most people in the world live in low-lying delta areas, coastal and river plains. This is not surprising, since fresh water for household use, irrigation or other outdoor purposes is close at hand and the land is usually quite fertile. The downside is the risk of flooding to which people are exposed. Throughout the ages various cultures have developed

Interview:

Water is linked to our identity

I live and work in a swampy delta area in Nigeria. The people living in the area do not have access to safe drinking water because of environmental pollution problems caused by the oil company's activities. In the delta areas there is a lot of spiritual, recreational and economic attachment to water. Traditionally, a newly born child is immersed in water, as their culture demands. Thus, if the water body is destroyed, the people will be denied this unique cultural attachment to it and their identity. The companies that directly exploit the natural resources deprive the people of their distinct cultural heritage because of economic gains. In some rural areas, people fetch water from streams for many uses. Drinking water is fetched in the morning because then the pollution is believed to have been washed away. In the afternoon one may do his or her laundry or take a bath in the stream. More importantly, when water is scarce, boys, deemed to be stronger, play a vital role in collecting water from distant and difficult places. Consequently, water distribution in itself is a kind of power struggle, since everyone tends to move towards the source at the same time.

Maria Assumpta Ngozi Anabi, Wastewater manager in Nigeria (2008).

[12]

different mechanisms of dealing with this risk. These include both successful and less successful strategies, as recent floods such as the disaster in New Orleans show.

The Dutch are renowned throughout the world for the way they have managed to live in the estuaries of the Rhine-Meuse delta. Half of The Netherlands, literally meaning 'low land', lies below sea level. In the past there were many catastrophic floods that washed away entire regions and killed many people. Such dramatic events made the people realize that human intervention was needed to save the land. Already around 1000 AD large monasteries and counts took the lead in building the first dikes and draining the land. Later, from 1300 onwards, specialized administrative bodies were formed, the so-called water boards, or in Dutch, *Waterschappen*.[13] These became responsible for regional and local water management, their main task being to ensure

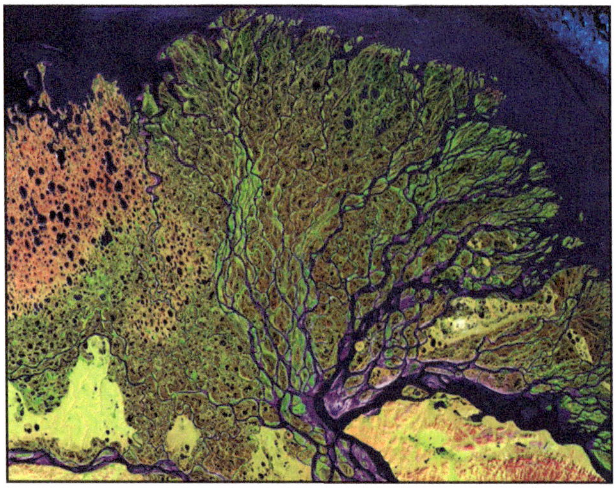

Figure 2.4 Lena Delta, R ussia.
Source: www.SEOS-project.eu

"dry feet" for the inhabitants. As a water authority they were and still are also legally allowed to collect fees for performing their tasks. As the years went by and with additional government funding, an extensive dike system was constructed in the Netherlands and land reclamation projects were undertaken, converting peat bogs and marshy areas, lakes and even inland seas into polders. The struggle against the water and the success in mastering it has played an important role in the development of the Netherlands and even in the development of the Dutch people's character. It is said about Dutch people that '*water is in their soul*'. This certainly applies to the people living in the low-lying Dutch province of Zeeland, who experienced a major flood in 1953, after which the world-famous Delta Works were built to protect them. Recently, the majority of Zeelanders strongly opposed new plans, prepared with the best of intentions by their fellow countrymen, to deliberately flood some polders to create more space for water.

At the same time, Dutch water engineers very well realize that the water protection system, which works so well for the Netherlands, is not the solution for all delta regions in the world. In many other delta regions, people have adapted their lifestyle, either urban or rural, to the flow and tides of the water. Take, for instance, the delta of Bangladesh, with its many freely meandering river systems. Let alone cost reasons, it is not feasible to build dikes around the entire estuary. Having no other choice, the inhabitants of this delta, many of them small-scale farmers, have adapted their way of life to the changing water conditions and 'move along with the water'. At times of high water they pick up their belongings and retreat to higher ground. After the floods they move back to farm the land, which has become fertile once again thanks to the silt and sediment deposits. Foreign aid to this country is therefore nowadays mainly directed toward better protecting the Bengali people in a dynamic and flexible way against flood events, for example by building retreats at still higher elevations, so as to sustain their way of living with water.[14]

2.3.4 The secret to survival in the desert...

... is knowing where to find water. People living in deserts, the most inhospitable and driest places on earth, are renowned for their ability to detect the presence of underground water. The movements of nomadic desert people are usually defined by the routes to water sources. Take, for instance, the vast desert in the United Arabic Emirates. Parts of the desert become green after the winter rain, and the news spreads quickly among the Bedouin nomadic people, who arrive in the area to settle until the hot and dry season arrives.[15] They have their own sophisticated methods of irrigation and tracing underground water using astronomical calculations. And they have mastered the technique of drilling accurate artesian wells and preserving these wells from sand storms and erosion. Water is brought up in a bucket and taken home in containers made from animal skin.

Speaking about animals, desert animals and plants themselves are also well adapted to living in this tough, water-scarce environment. They have developed internal mechanisms to survive extended periods of dehydration. And like humans, many animals, from small termites to big elephants, know exactly where to find the water. Sometimes too well... There is a well-known case in Zambia, where foreign experts had installed a new well and drinking water supply system for a local village community. They hadn't thought about the elephants, though, which were also thirsty and in search of water. Within a few days the elephants had localized the new underground water supply system and had damaged the pipes to drink from the water themselves. The only solution was to give them their own drinking water reservoir at a distance from the village.

Getting back to humans, there is ongoing debate about using precious water resources for irrigation purposes in desert areas unsuited to cultivation, where water evaporates quickly under the scorching sun.

Figure 2.5 Nomadic desert people in Afghanistan.

In arid countries like Afghanistan and Pakistan people have long overcome the problem of evaporation losses by making use of so-called **Karez or Qanat systems.**[16] These are ancient underground irrigation systems, owned and operated by the local community. The Karez system is comprised of a series of wells and underground water channels, relying on gravity to bring groundwater down from the mountains to the flatlands, often far from the source. The water is quite clean and can be used for irrigation as well as for providing drinking water to the local community. The water is taken to the fields and villages through vertical shafts which are sunk underground, or it is drawn out at the foot of the hill where it has been gathered. This ancient and social water supply system originated over 2000 years ago in the Chinese deserts west of the Himalayas. The construction, operation and maintenance of the Karez, including the cleaning of the system, are a collective social responsibility. In the early days large networks of these Karezes existed in countries like China and Pakistan. Many of them fell into disuse and decay, the reasons not always being quite clear. Similar systems can be found in other areas of the world, as in Northern Africa, Southern Europe (Spain) and even Latin America, where qanats can be found in the Atacama regions of Peru and Chile at Nazca and Pica. The Spanish introduced qanats into Mexico in 1520 AD.

Recently, many of the traditional Karez systems have been restored to use. In Afghanistan, the rebuilding work is part of a UN peace mission and involves restoring the existing Karez systems to their original state with the help of the local people.[17] This is a good example of **revival of old traditional practices.**

On yet another dry continent, Australian indigenous Aboriginal people were also able to survive in the desert thanks to their skills of digging underground water reservoirs and knowing where to find water. They had extensive knowledge of the subsoil groundwater system and also obtained water from certain trees and roots. Their local knowledge and sustainable use of water allowed them to live in the deserts, where others would perish, for many thousands of years.[18]

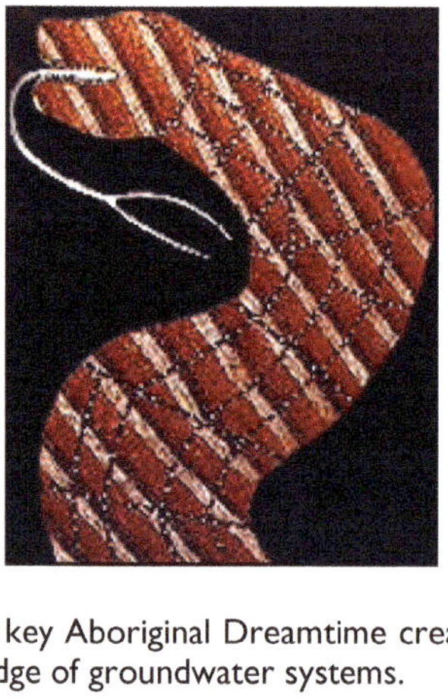

Figure 2.6 The rainbow serpent, a key Aboriginal Dreamtime creation symbol, is closely connected with indigenous knowledge of groundwater systems.
Source: Reuters

Interview

'We should build new water management systems on the successes of old traditions'

Yemen has a long-standing tradition of sustainable water management. Already thousands of years ago one of the earliest irrigation systems in the world was established. Civilization prospered, with a peak during the Kingdom of Sheba, thanks to well-functioning rainwater harvesting techniques. These included diversion dams, like the famous Marib Great Dam. Mountain terraces were vital in retaining rainwater for crops. Rainwater falls unpredictably in our country, but when it finally falls, it does so abundantly. In a semi-arid country like Yemen it is therefore vital to collect it. Moreover, most of our old cities were self-contained regarding sanitation by keeping solids and liquids separate. The liquid part was used for irrigating home gardens and the solid waste was used as an energy source for heating the water for public baths, etc.

But we abandoned the old systems instead of improving them and replaced these with modern technologies, such as large-scale extraction of groundwater. As a result, scarce groundwater sources are now drying up. In hindsight we can now say that we shouldn't have abandoned well-functioning traditional systems; modern technologies are not working. Nowadays there appears to be a lack of respect, especially amongst young people, for all traditions. They consider traditional knowledge as backward. Of course cultural traditions are not a panacea, but in order to regain a sustainable water management, we should try to build further on what has worked successfully in the past.

Mr. Abdul-Rahman F. Al-Eryani
Minister of Water & Environment
Republic of Yemen (2007)

2.4 THE CHALLENGE OF CHANGE

We live in a changing world. A recent EU publication entitled "Global trends affecting the water cycle: Winds of change in the world of water",[19] describes ten major trends that are expected to have widespread, significant effects on the world of water in the coming decennia. These include climate change, urbanization, globalization and energy use and costs. Knowing the trends that cause change allows for timely interventions. These may range from adaptive strategies involving change to fit new circumstances, to mitigation measures to counteract the trends and its impacts. As can be concluded from the previous section, humans have developed sophisticated survival mechanisms to adapt themselves to living with water under different circumstances. This section focuses on climate change and urbanization and the ways in which we can adapt our water management strategies to these yet irreversible trends.

2.4.1 Climate change

2.4.1.1 Severe water-related consequences

Thanks to Al Gore[20] and the important work of the International Panel on Climate Change (IPCC), we all are now aware that the earth's climate is gradually changing. Changes are already manifest everywhere around the globe. As a result of global warming, the polar ice caps and almost all of the mountain glaciers in the world are melting. The Himalayan glaciers on the Tibetan Plateau have been among the most affected by global warming. The river systems originating from this plateau are gradually drying up. As a result the drinking water supply of the over 500 million inhabitants of the region is at stake unless we are able to mitigate global warming in time or find other ways to foresee in the need of water.

2.4.1.2 Too much...

The weather is also becoming more **unpredictable**. Could farmers in West and East Africa fairly accurately predict the start and duration of the rainy season, in the last decennium there was no rain for many years in a row, with unexpected heavy rainfall in September 2007. The resulting flood zone stretched from the Atlantic coast to the Red Sea and had devastating effects on communities. Many lives were lost, homes were washed away and the planted crops for the coming year were totally destroyed. The flooding also ruined the supply of clean water, so both food and water have become scarcer than ever, with disastrous consequences lying ahead for the people living in these areas.

2.4.1.3 Too little...

On yet another continent, Australia has been experiencing severe drought problems, particularly in the Murray-Darling catchment area in the South and Southeastern part of Australia. As a result there is not enough water to supply both farmers and citizens and to preserve the precious river ecosystems. This causes tensions, as citizens accuse the farmers of using too much water for irrigation purposes, especially for the cultivation of rice and cotton, while farmers argue that they aren't the ones who should be made to suffer because citizens overuse water in their gardens. Nevertheless, plans were made to restrict the use of water for irrigation purposes in 2008 altogether. But the darkest hour is before the dawn; recent rainfall brought some relief, and although it wasn't enough to replenish the rivers and basins, the plans for usage restrictions were able to be set aside for the time being.

2.4.1.4 Too filthy...

The climate changes expected will also affect the quality of water resources. Higher water temperatures increase the likelihood of blooms of toxic algae and bacteria

Figure 2.7 Flooded land in West Africa, September 2007.

occurring in surface water. Furthermore, floods and sewer overflows due to more frequent storm events may cause deterioration of the water quality and health effects in deltas and urban areas.

2.4.1.5 ... or better?

Despite the adverse impacts, climate changes may also bring some positive changes, for instance in Greenland, where permanent frozen lakes are already melting. Despite the ecological consequences, this offers new socio-economic opportunities for the livelihoods of local fishermen and their families.

The Inuit are renowned for their ability to flourish in a harsh climate, to adapt as conditions change, to thrive where others cannot. Climate change poses a threat unlike any they have faced before. Their lifestyle and their culture may depend on their ability to adapt to this new challenge. The recently published documentary of Jan Louter, "The Last Days of Shismaref" (*http://www.imdb.com*, 2008, 90 mins) illustrates the severe consequences of climate change for local communities, in this case the coastal city of Shismaref in Alaska.[21]

2.4.2 Coping with climate change

2.4.2.1 Coping with unpredictability

A main challenge for water management is to be able to cope with the unpredictability of climate change and to avoid disasters resulting from more extreme weather conditions. Worldwide, countries have to deal with the effects of climate change and share similar dilemmas. These range from water shortage caused by prolonged, hotter dry periods to a water surplus due to the intensification of precipitation and flood events. The latter is exacerbated by the expected sea level rise, and people living in lowland delta areas will be the first ones to suffer from increased flood events.

2.4.2.2 Reading the signs

Water shortages are expected to become worse in semi-arid and arid regions such as Africa and Australia. In the Sahara and Sahel regions of Africa, desertification and the expansion of existing deserts, primarily caused by deforestation, over-exploitation and inappropriate land use, forces people to relocate to other regions and change their lifestyles. Scientists have lately discovered that sudden changes in vegetation patterns are an early warning signal for the occurrence of a sudden, catastrophic shift towards desert conditions. Being able to read the signs in vegetation patterns therefore provides a valuable tool for predicting desertification and allowing timely action.

Here again, there are positive signs, too. A most amazing one is the discovery made by a scientist returning to the Niger area after 10 years who expected to come back to the same desert and barren area that he had left. At the time, in the 1980s, the local farmers were suffering from crop failure and food shortages as a result of a severe drought. But to his surprise, the area had been transformed into a green, tree-covered landscape. Curious about what had caused this transformation, he asked the locals and learned that the turning point had been an encounter with an Australian missionary who had visited the area. He had made a simple deal with the local farmers: if you protect the trees, I provide you with food. This triggered the farmers to start protecting the trees that naturally grow in this area. They also made use of a simple traditional water harvesting method of digging 25 cm deep holes in the soil in which they put manure from their cattle and planted seeds. These holes also served as small water reservoirs. Soon new trees started growing out of them, which the farmers protected with their own lives. They were so dedicated because they saw the advantages of the trees: they shielded their crops from the desert wind, provided shade and food to their cattle, and the dead branches of the trees could be used as firewood. This proves that **solutions that are developed and maintained by the local people themselves stand the best chance of success.** In this case the transformation was so amazing that even the Nigerian government officials were sceptical until remote sensing images from the area convinced them.

In the coastal city of Semarang on the Indonesian island of Java, the residents of one poor community have become accustomed to having their houses flooded almost every day during the rainy season by seawater mixed with polluted water from the canals. Yet living under these conditions is by no means easy and is obviously very unhealthy. There are now serious plans along with a technical design to develop a polder with dikes and pumps to protect the area from flooding. A provisional Water Board*, in which the villagers actively participate, is already in place. The people living in the area are clinging to these signs of hope for a better living environment.

** The institutional component of the Banger Pilot Polder Project is part of the LOGO South programme of VNG International, implemented by Hoogheemraadschap van Schieland en Krimpenerwaard and the City of Semarang*

2.4.2.3 Need for a holistic and socially sustainable approach

In contrast to the previous success story, water management practices do sometimes also contribute to desertification. For a long time the foreign aid strategy towards water management in the Sahel regions was directed at digging more and deeper groundwater wells. This unfortunately brought only temporary relief for the nomadic people and their herds living in this area. At first, the people were pleased to have easier access to safe drinking water and the size of their stock of cattle increased dramatically. Roaming the area in search of food, however, the animals destroyed the scarce vegetation. In addition, subsequent years of hardly any rainfall made the wells dry up, leaving not enough water to quench the thirst of both humans and cattle and leading to massive starvation. Recent donor programs tend to take a more integrated approach, addressing socio-economic as well as environmental dimensions, such as reforestation programs. It is now broadly recognized that for water resource management to be sustainable a more holistic view is needed. And sustainability can only be reached when it includes the social aspect. This requires a more "people-centred" approach. Too often *the human factor* is, even with the best of intentions, forgotten. Using the Sahara region as an example, in many villages drinking water pumps were installed to save the women and girls the long and strenuous daily journey to fetch water. In most cases this was considered a welcome improvement, leaving the women with more time for household or outdoor activities and allowing the younger girls to spend more time at school. Yet in some Islamic communities this daily walk, even if it took several hours, was the only chance the women had to get out of the house and socialize with the other women. Water had an important social function for women, and they would have liked the water source to be a little closer but not *that* close to their homes...

2.4.3 Urbanization

2.4.3.1 The rapid pace of urbanization

In search of a better future and more economic prosperity, many people moved from an agricultural existence to urban cities during the last century. They left behind their old and often sustainable lifestyle, in tune with nature, to adapt themselves to city life. People living in cities also tend to adopt a more individualistic lifestyle compared to the high degree of collectivism found in rural areas. By now there are almost 400 cities in the world that have more than one million inhabitants and 18 cities with more than 10 million inhabitants, most of them in Asia. Today, half the world's population lives in urban centres, with around a billion urban residents living in slums[22] – and this is not the end, as more and more people are attracted by the modern joys and facilities of city life. Of course there are many downsides, too.

Apart from the problem of land subsidence and the resulting flooding in big delta cities, such as Bangkok and Jakarta, the ongoing trend of urbanization has large implications for freshwater use and for waste water treatment. The installation of water supply and sanitation facilities and the capacity of governments to manage them could not and still cannot keep up with the rapid pace of urbanization. Moreover, the expanding cities must draw fresh water from increasingly distant water sheds, as local sources become depleted or polluted by direct (untreated) discharges. In a semi-arid country like Spain, additional water stress is caused by the many newly developed golf courses requiring huge amounts of water to keep the greens green. In summertime, with soaring temperatures and little to no rainfall, large amounts of tourists invade the coastal cities and villages of Spain, leading to an actual shortage of drinking and irrigation water. The city of Barcelona has recently tried to overcome this by supplying potable water by ship. It is also working on other solutions, such as seawater desalination. These options cost about €30/m^3 (2008), which is 15 times the cost of the regular water supply.

2.4.3.2 Water for food

Shortage of water also causes increased tension between urban and agricultural demands, as was illustrated earlier in the Australian example. On a global scale, population growth, especially in countries like China and India, leads to an increased need for food production. The high demand not only drives up the market prices of food, but also leads to more water consumption to irrigate crops and breed livestock. Worldwide initiatives are being taken to solve this problem by improving irrigation practices ('more crop per drop'), raising and selecting crops for which the water requirements match the availability of water ('function follows water level') and promoting less consumption of meat, as its production requires far more water than is required for a vegetarian diet. According to the United States Department of Agriculture, raising crops for farm animals requires nearly half of the US water supply and 80% of its agricultural land.

Another serious problem associated with increased food production is that of increased use of slowly degradable pesticides. They usually end up in the water cycle, causing pollution of water bodies and thus forming a threat to humans by direct contact or through the consumption of water or fish. Locally, small-scale initiatives are being taken to produce organic products. This problem deserves more attention on a global scale.

2.4.3.3 Situation in slums

In many major Asian, African and South American cities, the poorest people live in slums, deprived of all comfort and the basic necessities to lead a humane life.

The day-to-day burdens of people living in slums are huge. Most of them have to fetch their own water, with bottled water being the only reliable – but expensive and often unaffordable – alternative. Local entrepreneurs are trying to fill the gap by installing water piping systems with payment meters in the "houses", but these water sources are often far from reliable.

In the urban slum of Kebira in Nairobi, Kenya, for example, home to 700,000(!) people, the water supply piping systems often run through waste water gullies. As the pipes themselves are usually fit together by hand, they tend to leak and take up filthy water from the gullies. People know they have to cook the water, but the water itself and the gullies especially are a breeding ground for bacteria, harmful insects, etc. and as such pose a serious health risk. Many children suffer from water-related diseases such as cholera and diarrhoea.

The poor or rather lacking sanitation facilities add to the problems. Most people use plastic bags to defecate, which are mostly thrown away randomly, creating not only a health risk but the habit also forms a nuisance and source of dispute within the community. Especially for women, the lack of adequate sanitation facilities has negatively affected their dignity.

Interview:

A difference between urban and rural water culture

Living and working as a water supply engineer in the big and quite "Western" Colombian city of Cali, I can think of no explicit cultural beliefs influencing water management.

In more inland, rural areas, culture plays a more prominent role. Having worked with indigenous people, I learned that they have a strong resentment to using a source of water that might have been polluted by others living upstream. They don't want to drink someone else's 'pee' as they put it. Groundwater was okay for them to use, even though technically speaking it was sometimes of lower quality. In rural areas there is also the strong belief that water should be available for free, as it always had been. People got very upset when we wanted to install water meters inside their homes. Their main fear was that once they got this water meter they would have to pay a high tariff for water, like in the city. In the end we managed to assure them that they were just paying a small and fair amount for the services delivered.

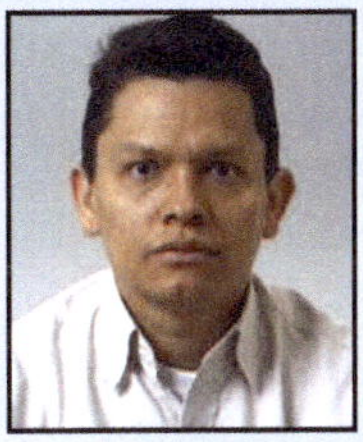

Arlex Sanchez Torres, Colombia

Kebira is not a stand-alone problem, nor is it unique in its sort. In many African, Asian and South American urban slums similar problems occur. Slowly, reliable and robust water supply, sanitation and waste water treatment systems are being installed – too slowly, really, as still more than a billion people in the world, living both in urban and rural areas, have no access to safe drinking water, and two billion have inadequate sanitation. Efforts to meet the Millennium Development Goals by reducing these numbers by 50% before the year 2015, are seriously hampered by further urbanization. There is also a clear relation between poverty, economic growth, health issues and education. It has been calculated that every dollar invested in drinking water and sanitation in developing countries has a seven-fold return on investment.[23]

Many successful programmes have demonstrated that water and sanitation provisions benefit from strong community participation.

2.4.4 Water management in cities: the need for flexible and robust solutions

2.4.4.1 A different portfolio of options

An emerging challenge to urban communities is its design for **resilience** to climate change. Flexibility and robustness are important in this respect, as they enhance resilience. In order to cope successfully with the effects of climate change and rising water demands, modern urban communities have to be adaptable and flexible by having a different portfolio of options.[24] Adaptive strategies for water supply may include having separate systems in place for potable water, for industrial use and waste water re-use. In Australia most major cities along the coast nowadays rely on a diversity of sources and infrastructure, allowing them to have access to storm water in a wet year and access to desalinized water in a dry year. Public involvement in such plans may enhance their acceptance. However, plans in the drought-stricken town of Toowoomba in the Australian desert to produce drinking water from treated waste water failed because of fierce opposition from the local population. Re-using water from "toilet to tap" was clearly a bridge too far here, as it was not in line with local notions.

2.4.4.2 Integration of spatial planning and water management

Of a different category are adaptive strategies aimed at the integration of spatial planning and water management. There are still too many home development projects in the world today that are based on economic, social or architectural concerns rather than on a choice for environmental sustainability. The result is too much built-up, non-porous surface area with too few green spaces. Another problem is homes being built – often uncontrolled and illegally – in dammed river or flood plains, leaving the water nowhere to go other than to flood these homes one day. A example of this is the overcrowded city of Jakarta, on the Indonesian island Java, which suffered a major flood in the month of February in the two consecutive years of 2007 and 2008. In both cases heavy tropical rains had caused rivers to overflow their banks, sending muddy and filthy water up to 2 metres deep into people's homes and businesses, leaving almost half a million people in Jakarta homeless and devastated. Some people drowned or were immediately killed by electrocution. Many others suffered from outbreaks of diseases like dengue fever, diarrhoea and leptospirosus, caused by rats urinating in the water. As always, the poor got the worst end and suffered the most, yet in the end it always affects and concerns everyone. In this case several "richer" communities were also affected, their homes being flooded due to sabotage of the pumps or protective walls, resulting from the social jealousy of neighbouring poorer communities ('If we suffer, you should suffer as well').

Anyone who has ever witnessed a major flood and its aftermath recognizes the urgent need for integrated spatial water management to successfully achieve

Figure 2.8 Flooded densily populated area of the city at Jakarta, Indonesia, February 2008.

sustainable development along with pollution control, flood control, drainage and clean water supply.

2.4.4.3 *Early warning systems*

In fact, the entire world witnessed the *tsunami* on December 26, 2004.Triggered by a huge earthquake under the Indian Ocean, giant waves were sent up to the shores of coastal areas across South and East Asia, killing at least 200,000 people and leaving millions homeless and devastated. Not surprisingly, poor people living in frail shelters on marginal lands in coastal plains were affected the most. It also took longest for them to rebuild their lives, despite the vast international relief effort. Under the auspices of the United Nations, long-term measures were also taken to prevent disasters like these in the future. Unlike the Pacific, the Indian Ocean did not have an early warning system in place to alert residents of coastal areas that a tsunami was imminent. Within a year after the tsunami the first of a network of early warning buoys were installed in the Indian Ocean.

Extreme New Orleans: a haphazard response

Poor decisions made throughout its 288-year history made New Orleans so vulnerable to hurricane Katrina in August 2005. These reflect a long-term pattern of ad-hoc societal response to previous hazard events. The measures taken reduced the consequences of small, relatively frequent events but thereby increased the city's vulnerability to rare, extreme events. Katrina's consequences for New Orleans were truly catastrophic, accounting for most of the estimated 1570 deaths of Louisiana residents and 40-50 billion dollars in monetary losses. In its 288-year history, New Orleans has had 27 major river or hurricane-induced disasters, at an average rate of one every 11 years. After each event the city rebuilt and often expanded. Small differences in elevation determined the residential locations of the well-to-do and the poor. The response to riverine and hurricane-induced floods was to build levees. Already by 1728 it was mandatory for all land owners to build these along their riparian frontage. This responsibility gradually shifted from land owners to the state and ultimately to the federal government. The political culture in the 20th century was such that development was often rewarded at the expense of safety. After hurricane Betsy (1965) the US Army Corps of Engineers prepared a major plan to protect New Orleans and its surrounding areas from hurricanes and flooding. Conflicts between local and federal authorities over the final form of the hurricane protection system greatly delayed its completion and exposed everyone to heightened risk. There was also opposition from the predominantly Afro-American population to any proposals coming from the local authorities. Yet the distinctive role New Orleans plays in African-American politics, culture and education was also a main reason to rebuild the city. This provides a chance to rebuild New Orleans in a sustainable way, thereby properly addressing the issues of race and poverty that are deeply embedded in the society. Extreme events, like Katrina, reveal the extreme differences in the way we live and die, cope and rebuild.

Sources: Prof. Craig Colten, Department of Geography and Anthropology, Louisiana State University et. al., *Reconstruction of New Orleans after Hurricane Katrina: a research perspective* (Oct. 2006);
and form:
Lecture by Prof. Craig Colton in Lelystad,
The Netherlands, September 2006:
'Extreme New Orleans: Building Beyond the City's limits'.

Not much later, in August 2005, the world and especially the United States were shocked by the damage caused in New Orleans by Hurricane Katrina. Despite prior warnings, many people had refused to leave town and were killed or trapped inside the city. A conflict of interests and cultures contributed to the major impact Katrina had. Previous plans made by the American Corps of Engineers to safeguard New Orleans and its surrounding town from floods caused by hurricanes were never realized because of opposition from local people who mistrusted any plans from the local authorities, environmentalists, project developers and conflicts between authorities.

2.4.4.4 New concepts, building on old traditions

Prevention is always better than a cure, and in this respect flood-resistant building methods also offer opportunities. The idea is not new. In the United States and many other countries, building homes to resist flooding traditionally meant raising them and using the basement to accommodate periodic flooding. In many coastal areas in the world people live in houses built on stilts over the sea. Dutch architects and project developers have taken the idea further. They are actually designing floating cities and buildings that can rise or fall with the water level. There is a world-wide interest in these designs.

Currently a trend is also visible towards self-sustaining, eco-designed homes and office buildings. These provide, for instance, their own electricity from solar, wind or geothermal power and harvest rainwater for domestic or garden irrigation use by collecting it from the roof. Again, this is not quite new, but builds on old cultural practices. In fact, most of our parents and grandparents lived in a time when houses were not connected to a central drinking water supply system, but rather were equipped individually with a storage container for water, in which water falling from the roof was collected. In coastal fishing villages in the Netherlands, the water was first purified by flowing through a container of shelves before falling down into the storage reservoir. The water in the reservoir was kept in movement by a fish (usually a stilt) swimming in it, to prevent algae blooms. Nowadays we would perhaps use an electric – or even better, solar-driven – agitation device, but back in those days this technology was not available. People also didn't use as much water as we do today, and in this respect we can learn a lot from our ancestors about water saving practices.

2.4.4.5 Balance between wet and dry seasons

In countries where there is a surplus of rain during the wet season and a shortage during the dry season it is a wise strategy to find a balance between the wet and dry season. This can be done by harvesting the rainwater when it falls and storing it for use in times of shortage. The famous Ghandi was ahead of his time in this respect, as the family home he lived in had a huge underground storage basin for rainwater. Each year before the monsoon season started, the roof was cleaned and the water falling abundantly from the sky was collected and led to this storage facility in the basement of his home. This was sufficient to supply the family's water needs year-round. Such solutions will become more important as urbanization, together with increased population growth, will cause even more severe water stress world-wide. Ecological solutions like reforestation programs offer further perspectives, even within cities, as they increase the retention capacity of the soil and improve urban living conditions.

In the face of climate change it is important to enhance the natural resilience of the system. This implies the ability of the system (both ecological and social) to rebound and withstand changes in the water supply induced by climate change. It involves the inherent ability of ecosystems and communities to adjust to new circumstances.

2.5 CONCLUSIONS

Key question: Can we incorporate the "old" proven adaptation mechanisms into new sustainable water management strategies?

As a survival mechanism, human societies have always adapted their lifestyles to be in balance with their environment. Traditional societies still teach us how to ensure ecosystem resilience and cope with the challenges posed by the natural environment and to use its resources without disturbing the delicate balance.

We can learn from the ways people used to deal with a shortage or abundance of water in different parts of the world and re-introduce proven practices to achieve more sustainable water management in our modern societies. Of course, traditional practices are not a panacea able to solve all water-related problems in present-day times. Sustainable water management will therefore often mean finding the right mix between "old" and "new" practices. In dry climates this may be a combination of practices like harvesting rainwater, minimizing evaporation losses, improving leaky systems, introducing more efficient irrigation practices and choosing crops that require less water. Dry sanitation systems that need no water for toilet flushing and which keep the liquid and solid waste separate for re-use offer opportunities in places where water is scarce. In flood-prone areas this may involve improving present flood control structures to be able to handle more frequent and extreme events, as well as enhancing the natural capacity and resilience of the water system to accommodate water. It also entails developing more advanced forecasting and early warning systems along with evacuation plans to allow people to leave the flood-prone area in time.

Key messages

- Water is a critical element for sustaining life.
- In many traditional cultures water is linked to people's identity; if you destroy their water resource, you deprive them of their cultural identity.
- In many cultures water has an important social function, especially for women.
- Professionals can learn from the ways that cultures around the world have adapted themselves to living with water in different environments.
- New water management systems can be built on the successes of old traditions.
- Solutions that are developed and maintained by the local people themselves stand the best chance of long-term success.
- It is important to acknowledge and enhance the natural resilience of the system.
- Public participation is key to successful and sustainable water management.
- A sustainable approach includes finding a balance between water surplus and water shortage.

3 – WATER: A SOURCE OF INSPIRATION

Chapter 3

Water: a source of inspiration

> The shining water that moves in the streams and rivers is not just water, but the blood of our ancestors. If we sell you our land, you must remember that it is sacred. Each glossy reflection in the clear waters of the lakes tells of events and memories in the life of my people.
>
> The water's murmur is the voice of my father's father. The rivers are our brothers. They quench our thirst. They carry our canoes and feed our children. So you must give the rivers the kindness that you would give any brother.
>
> (Earth wisdom of Suquamish Chief Seathl, 1854)[25]

3.1 INTRODUCTION

As one of the crucial life-sustaining elements, it is only natural that water has given rise to a multitude of beliefs, both religious and spiritual. These range from ancient mythologies and the fundamental nature beliefs of indigenous societies to the contemporary teachings of the major world religions. One common thread running through all of these is a reverence for water. But paradoxically enough, despite the deep respect for water and its prominent place in cultural and religious beliefs, in everyday life water is often taken for granted, polluted, spilled and fought over. How is it possible that in many societies with so much water wisdom this hasn't resulted in wise water management but rather the opposite? And can religion still become an alley in achieving sufficient, safe and clean water for all? Let's find out...

We'll take a cultural and religious journey across the continents and the main world religions, starting in this chapter with the oldest known form of human religion, also referred to as animism or nature religions.

The next sections (3.2–3.7) deal in chronological order with the way Hinduism, Buddhism, Judaism, Christianity and Islam view and treat water. This is subsequently followed by a section (3.8) about other contemporary religious beliefs and world views, in particular Taoism, Baha'i and the 2012 movement. The chapter ends with a concluding section (3.9) on wise water management.

3.2 WATER AND ANIMISM

Animism is characteristic of aboriginal and native or indigenous cultures but continues to exist today, even in modern societies and in co-existence with contemporary world religions. Animistic beliefs are still prominent in Africa, parts of South America and Asia, especially in Indonesia and the Philippines. The Water Page[26] gives an elaborate and interesting description of the relationship between animism and water. Much of the information in the following overview of water's role in animism has been derived from this source.

Animism is often defined as the *'belief that humans, natural objects, natural phenomena and the universe itself possess souls that may exist apart from their material bodies'*. Some adherents of animistic religions revere specific water spirits, while others believe that water itself possesses supernatural qualities. The manner in which this life force is visualized is dependent upon the specific beliefs of the people in relation to the environment they inhabit. The physical forms of the water spirits in which people believe range from serpentine to human or mermaid forms, including a variety of intermediate forms. The serpentine form tends to dominate in Western and Southern Africa, whereas the half or wholly human mermaid form prevails in Northern and Central Europe.

Animistic beliefs and the water spirits associated with them travelled along with the movement of peoples across vast distances and from continent to continent. The Atlantic slave trade running between West Africa and the colonies of North America not only transported enslaved people but their belief systems as well. As a consequence, water spirits almost human in form and known as "cymbees" inhabited springs in the low-lying areas of South Carolina. They were very similar to the "simbi" water spirits feared by the indigenous peoples of the Congo basin. In South Carolina they were mainly feared for their power to stir up very high winds and unleash tornadoes, a reoccurring and frightful phenomenon in that region.

Malignant water spirits, whether at sea or in lakes and rivers, are believed to be the cause of all kinds of evil related to water, such as drowning. In parts of Poland and Germany people used to believe in the waterman or *nix*, who possessed a human form and a malevolent nature. Inhabiting lakes, rivers and ponds, he tempted passers-by to go bathing in order to drown them. In the Baltic region, local fishermen believed that some of the spirits could make their boats lose their way or prevent them from catching any fish. Offerings of a first catch to appease the spirits were commonplace.

In the same Baltic area, water spirits were believed to be able to cause illness. By making offerings to the spirits and asking for forgiveness at the springs or streams, one could be cured. The belief in the healing power of the water spirits was not limited to the Baltic region; it was a common phenomenon across much of Europe. However, with the introduction of Christianity, the springs became rededicated to the cult or veneration of the saints. In some cases, where the indigenous belief was strong, the Church simply made the god or goddess a saint and assimilated the local belief. Those suffering from illness still came to the spring for healing, although the credit went instead to a Church-sanctioned saint. There is much more evidence that in fact all nations held wells and fountains in a kind of religious awe. Even today there are many contemporary springs to which people go in hope of cure, Lourdes in France being one of them.

Aside from the presence of water spirits, the water itself is often believed to possess supernatural properties, such as healing, harmful or protective qualities. The belief in these qualities can exist alongside or be incorporated into the major religions of Christianity or Islam, illustrating the resilient nature of animism. An example is the longstanding tradition in Germany of using holy water to protect oneself and

A well-publicised example of a water spirit is that of the Nyaminyami of the Tonga people. This spirit is believed to inhabit the Zambezi River between Zimbabwe and Zambia. The Tonga (who lived on both banks of the river before their forced removal with the construction of the Kariba dam) regarded him as a god. Although only a few sightings have been claimed, his physical form is serpentine, with a snake's body and a head similar to that of a tiger fish. In times of hunger, he acted as a protector toward the Tonga, giving them sustenance by providing strips of meat from his own body. In return the Tonga demonstrated their allegiance with ceremonial dances in his honour. Nyaminyami had a wife, and together they roamed between Kariba, the Kariwa gorge and the Mana pools. However, he was separated from his wife by the building of the dam at Kariba.

During the construction of the dam in 1957, the Tonga people were forcibly resettled from the banks of the Zambezi to the surrounding barren highland areas. However, construction was set back by the occurrence of a millennial flood. The resulting damage was the destruction of the constructional coffer dam. Following floods proceeded to remove the suspension footbridge and road bridge between Zambia and Zimbabwe. Further setbacks occurred, including the death of eighteen workers who fell to their deaths during construction. Nyaminyami was claimed to have been involved for two reasons. He was said to be lonely, as he was separated from his wife who was still residing at Mana pools, and so in his anger had caused the floods. The Tonga people also claimed that he had acted in defence of them when they invoked his protection as an act of resistance against their forced removal. Yet the Tonga also claim that the only reason that Nyaminyami did not completely destroy the Kariba project was due to the intervention of their elders to placate him so as to spare further destruction. Nevertheless, the completion of the dam and the resettlement of the Tonga people away from Kariba has not destroyed the belief in Nyaminyami. Occasional earth tremors are felt in the region. These are believed to be caused by the wandering of Nyaminyami, lonely and still wishing to be reunited with his wife. This, as the belief further goes, will eventually be accomplished by the destruction of the dam.

Source (26), Water page

one's goods and belongings. On receiving holy water from a priest, one could hang it around the neck in an amulet to ward off evil spirits or put it in a small basin above the door of the house to keep them from entering. These ancient established practices were adopted by Christianity.

In African communities, dependent for their livelihood on rainfall, animism allowed for a process of rainmaking with the use of "medicines" in times of drought. Modjadji, the Rain Queen of the Lobedu Mountains in South Africa, was famous for her rainmaking abilities. Even the members of the Boer republic of the Transvaal, considered the enemy of the local Bolebedu people, visited her in fear of being denied water from above. As such, her attributed abilities gave her kingdom greater political power than it could otherwise have expected.

A characteristic trait of animistic and indigenous water beliefs is that water is often personified: Water knows, hears, smells, protects and can be sorry or happy. Water is viewed as a living entity and could be a father, mother or brother, as in the famous words of Chief Seathl at the beginning of this chapter. The North American Indians also believe that the rivers and lakes contain water spirits. They therefore treat water with the utmost respect. Their philosophy of life is to use nature and water respectfully, never taking more than is needed. In a similar way, the Anishinabe

Indian people, living around the Great Lakes in Canada, think of rivers as the veins of Mother Earth, carrying her life-blood to all her children. Many are deeply concerned that her life-blood is being polluted, contaminated and depleted. In the same context of water personification, an Australian aboriginal person would ask another person: (What is your name? What is your mother's name? And from what waters do you come?) Their very identity is linked to water.

3.2.1 Indigenous people and environmental views

Indigenous people themselves can best explain the relationship they have with water. Stemming from a general feeling that indigenous cultural and spiritual understandings about water are misunderstood or simply ignored by the dominant Western societies, indigenous people from all around the world issued an official declaration at the 3rd World Water Forum in Kyoto, Japan in March 2003.[27] With it they affirmed the comprehensive and fundamental nature of their relationship with water, as expressed in their introductory words:

- *We, the Indigenous Peoples from all parts of the world assembled here, reaffirm our relationship to Mother Earth and responsibility to future generations to raise our voices in solidarity to speak for the protection of water. We were placed in a sacred manner on this earth, each in our own sacred and traditional lands and territories, to care for all of creation and to care for water.*
- *We recognize, honour and respect water as sacred and sustaining all life. Our traditional knowledge, laws and ways of life teach us to be responsible in caring for this sacred gift that connects all life.*
- *Our relationship with our lands, territories and water is the fundamental physical cultural and spiritual basis for our existence. This relationship to our Mother Earth requires us to conserve our fresh waters and oceans for the survival of present and future generations. We assert our role as caretakers with rights and responsibilities to defend and ensure the protection, availability and purity of water…*
- *… Our traditional practices are dynamically regulated systems. They are based on natural and spiritual laws, ensuring sustainable use through traditional resource conservation. Long-tenured and place-based traditional knowledge of the environment is extremely valuable and has been proven to be valid and effective…*

What relationship do the world religions have with water?

Let us begin in Asia. The religious traditions of Asia are rich and varied, offering diverse theological and practical perspectives on water.

3.3 WATER AND HINDUISM

3.3.1 Origin

Having originated some 5,000 years ago in the Indian subcontinent, Hinduism is considered the world's oldest major religion that is still practiced today. It is also the world's third largest religion, with approximately a billion adherents, of whom the majority lives in India and Nepal. Other countries with large Hindu populations include Bangladesh, Sri Lanka, Pakistan, Indonesia, Malaysia, Mauritius, Fiji, Surinam, Guyana and Trinidad and Tobago.[28] The word 'Hindu' has everything to do with water, as it is derived from 'Sindhu', a Sanskrit word for the Indus, the longest river in the Indian subcontinent. Sindhu was also the name used to refer to the people who inhabited the Indus banks in Pakistan and parts of India and Afghanistan as long ago as 2,500 BC. The roots of Hinduism date back to this peaceful Indus "river" culture.[29] Their peaceful life was disturbed around 1,800 BC when Indo-German nomadic tribes invaded their land. These brought with them their own gods, whom they were used to worshipping in "free nature". They did, however, incorporate the gods of the Indus culture into their own religion. This is a typical feature of Hinduism, which consists of a multitude of beliefs, with monotheistic and polytheistic beliefs in harmonious coexistence.

3.3.2 Role of water and rivers

Water in Hinduism has a special place because it is believed to have spiritually cleansing powers. Rivers have always been an integral part of Hindu religious practice. To Hindu people all water is sacred, especially rivers. The Ganges River is considered the most sacred of these. Flowing through northern India, it is referred to as a goddess (Ganga) originating from the top of Shiva's head in the Himalaya Mountains, giving sustenance to hundreds of millions of modern Indians.[30] According to Indian myths, the Ganges flowed from the heavens and purified the people of India who touched her. A notable feature in Hindu religious ritual is the separation between purity and pollution. Ritual impurities can be overcome by cleansing ceremonies, such as bathing in sacred water. Traditionally, the rivers of India have always been considered pure. Modern industrial contaminants and human waste, however, have seriously polluted the rivers.

Nonetheless, Ganges water still plays an important role in India's ritual life. One such ritual is the morning cleansing with river water, being a basic obligation for Hindu people. Hindu belief holds that bathing in the river causes the forgiveness of sins and likewise that immersion of the ashes of the dead in the Ganges will send the departed soul to heaven. Hinduism knows many sacred places. Crossings between land and water – or even better, between two or more rivers – are often sacred places.

The famous city of Varanasi, situated along the banks of the Ganges, is an important place of worship for Hindu people as well as a cremation ground. Pilgrims to this city bring offerings to appease the gods, and they often carry sacred water from the river back home. Visitors, witnessing the daily activities in the Ganges, are usually appalled when they see people using the water to bathe or brush their teeth while at the same time noticing dead corpses floating past. Religious traditions obviously prevail over what would generally be considered healthy practice.

3.3.3 Cosmic and environmental view

Hindu people belief that the cosmos is constantly changing, and therefore their holy scripts, the Vedas, need to be continually reinterpreted. The Forum on Religion and Ecology[30] provides a noteworthy description of how the broader values of Hindu

Figure 3.1 Indian woman praying at the Ganges River (Photo: UNESCO).

tradition might contribute to fostering greater care for the earth, including water as a valuable resource. They state that Hinduism offers a variety of cosmological views that may or may not situate humans in the natural world in an ecologically friendly manner. Most of the Hindu population lives within villages that barring natural disasters such as flood or drought are self-sustaining and use resources sparingly. Their lifestyle is seemingly in tune with the elements. However, as the population of South Asia increases, and as the modern lifestyle continues to demand consumer goods, the balance of sustainability could shatter. This notion, supported by the worsening air quality in South Asian cities and degradation of water in its regions, has prompted religious thinkers and activists to begin to reflect on how the Hindu tradition can develop new modalities for caring for the earth and its water!

Noteworthy here is the Swadhyaya Movement, which is based on the rediscovery of Indian Hindu religious and philosophical tradition for its relevance to contemporary society.[31] It currently has millions of followers across India and among the Indian diaspora abroad. Swadhyaya enshrines the notion of a universal religion that crosses and supersedes all religious barriers. An important feature of a person's relation with God is *bhakti* (devotion), which should be performed through active participation in community work on behalf of community welfare and human dignity. A person using his or her talent or proficiency for the community benefit, especially for the poor and the weak, is thought to offer his efficiency to God. Following this principle, Swadhyaya devotees took it upon them to voluntarily perform the required manual labour to recharge thousands of groundwater wells in several Indian regions.

3.4 WATER AND BUDDHISM

3.4.1 Origin

Buddhism is considered not only a religion but a way of life. Buddhism began around the 5th century BC on the Indian borders of present-day Nepal with the teachings of Siddharta Gautama, commonly referred to as "the Buddha". Born as a prince, he was confined to live a life of luxury within the palace gates, shielded by his father the King from religious teachings or human sufferings, which were omnipresent beyond the palace gate. The first disturbing things he saw when he met the outside world were an old man, a diseased man, a decaying corpse and an ascetic.[32] These encounters had

Figure 3.2 A Buddhist temple.

a profound influence on him and he decided to change his lifestyle radically. At the age of 29 he left the palace and his family. Through asceticism and concentrating on meditation he became enlightened and discovered the four Noble Truths of suffering and the Noble Eightfold Path that could end suffering: right understanding, right thought, right speech, right action, right livelihood, right effort, right mindfulness and right concentration. These noble truths and the path leading to them come under the heading of "Dharma" and form an important aspect of Buddhist teachings.

Buddhism was able to spread from India to China along the Silk Route, which stretched from the Mediterranean through Arabia, Egypt, Persia and India well into China. As Buddhism spread along this important trade and cultural exchange route it underwent profound changes. Reaching people of different civilizations, the demand for adapting the teachings to local customs shaped Buddhism in different ways. In this way in Tibet a unique form of Buddhism, mixed with the pre-existing animism, arose in the ninth century AD.[29] In Japan Zen Buddhism developed, which had a great impact on everyday Japanese life. Well-known is the Japanese tea ceremony; this is a religious Zen ritual. With estimates of 300 to 500 million followers, Buddhism is presently the fourth largest religion of the world, after Christianity, Islam and Hinduism.

3.4.2 Role of water

For Buddhists water is said to symbolize purity, clarity and calmness. As Buddhists seek spiritual enlightenment through meditation, rites are basically absent from this religion. Buddhists do use water during funeral services. Monks fill a bowl, placed before the dead body, with water to overflowing while simultaneously reciting '*As the rains fill the rivers and overflow into the ocean, so likewise may what is given here reach the departed.*'[33]

Water also plays a role in offerings, which are quite common in the East.[34] Related to Buddhist belief are the offerings of water to cleanse the mouth or face, signifying auspiciousness, and the offerings of water to wash the feet. The symbolic meaning here is purification. By washing the feet the negative karma is believed to be cleansed away. Though the term "holy water" is not used, the idea of blessed water is also used among Buddhists. Water is put into a new vessel and kept for protection after it is blessed.

3.4.3 Environmental view

As to the perspectives Buddhism offers on water management, these can be related to the Buddhist teachings on environmental concern. An important premise is the Buddhist notion that all beings are interconnected and exist by their interrelationship with all other parts of nature. It is therefore crucial to Buddhists to live in harmony with the environment. This is complemented by a great respect for the environment and for life itself. In the words of the Cambodian Buddhist monk Maha Ghosananda:[35]

> *When we respect the environment, then nature will be good to us. When our hearts are good, then the sky will be good to us. The trees are like our mother and father, they feed us, nourish us, and provide us with everything: the fruit, leaves, the branches, the trunk. They give us food and satisfy many of our needs. So we spread the Dharma (truth) of protecting ourselves and protecting our environment, which is the Dharma of the Buddha.*

Buddha himself taught people to appreciate the natural cycle of life and to not spoil, destroy or waste. He was also a strong advocate and a living example of following a life of simplicity and moderation. This entails not taking more than you need. According to Buddhism the demands for material possessions can never be satisfied; people will always demand more, thus threatening the environment. Care for a better environment therefore starts with the individual, and greed upsets the vital balance of life.

These days Buddhist monks play an important social role in local communities by raising the ecological awareness of local people and guiding them in protecting and improving the environment. An example is the Buddhist country Cambodia, where the people suffered for years under the Khmer Rouge regime and where the environment is now declining due to deforestation and poor waste management. The Association of Buddhists for the Environment now plays an important role in preserving the natural resources such as forestry, wildlife and aquatic resources.[36] The pagoda, traditionally being the centre of community life and learning, is re-emerging as an important centre of village life and environmental education.

3.5 WATER AND JUDAISM

3.5.1 Origin

Continuing the religious journey, we arrive at the Arabian Peninsula, presently known as Saudi Arabia and one of the driest places on earth. The Arabian Peninsula was the

Figure 3.3 A synagogue in Prague, Czech Republic.

cradle of three world religions: Judaism, Christianity and Islam. Judaism is the oldest of these three monotheistic religions, all still in practice today. It is also the world's first book-based religion, with many references to water in the Hebrew Bible. The history of Judaism begins with the covenant between God and Abraham, some 2,000 years BC. Today Judaism is practiced by over 13 million Jewish people in Israel and abroad. God's laws and commandments are captured in the Torah and laid down in the Hebrew Bible, or *Tanakh*. They are further explored and explained in the Talmud.

3.5.2 Role of water

Water plays an important role in ritual cleansing practices, having their origins in the Torah. Jewish people use water for ritual cleansing or "ablutions" to restore or maintain a state of ritual purity.[37] This involves the obligatory washing of hands before and after meals and on many other occasions. In the old days priests had to wash both their hands and feet before taking part in Temple services. Ritual baths, known in Hebrew as *mikveh*, played an important role in Jewish communities, but somewhat less so nowadays. A mikveh bath was taken as part of the initiation rites when someone was converted to Judaism. They are still compulsory for converts today. The mikveh has its origins in ancient times when people had to be purified in a mikveh before they could enter the Temple area. Over time the word mikveh, which in Hebrew literally means "a collection or gathering together", became associated with a collection of water, such as a reservoir or pond. It came to be that the mikveh had to be located at a source of natural water, as the water used for taking the baths must be "living (running) water", from the sea, a river or spring. Men were advised to take a bath before the Sabbath or before important festivities. Total immersion was also required after contact with a dead body and for women before their wedding, after giving birth and after menstruation. According to the Torah, a person could become unclean on such occasions, and by taking a mikveh bath a status of ritual purity was regained.

3.5.3 Water in the Bible

In the Bible water is mentioned at various times as both a punishment and blessing of God. First, there is the well-known story of the Great Flood and the ark of Noah. Genesis 6–9 tells how God decided to destroy humanity by sending a great flood. Only Noah and his family and pairs of all the animals and birds were saved, as God instructed Noah to build an ark. After the waters had abated, God entered into a covenant with Noah and his descendants. God promised never to attempt to destroy the earth again and sent a rainbow as a sign of this covenant.

The story of the Great Flood shares many features with the ancient mythology of the region. The earliest extant flood myth is the Sumerian Genesis.[38] It dates back to the Babylonian period in Mesopotamia around 1700 BC. In Babylonian mythology a man named Utnapishtim tells how his chief God warned him of the gods' plan to destroy all life through a great flood and instructed him to build a vessel in which he could save his family, his friends, and his wealth and cattle. After the Deluge the gods repented their action and made Utnapishtim immortal. In the Babylonian story the destruction of the flood was the result of a disagreement among the gods; in Genesis it resulted from the moral corruption of humans.

In fact, the story of a Great Flood or Deluge sent by a deity or deities to destroy mankind as a kind of retribution appears in many other religions (Hindu, Christianity, Islam), in cultural myths and stories around the globe, from India to South America, and in different time periods. These may be attributed to actual local or large-scale flooding events, as there is ample geological evidence of tsunamis, sea-level rises, deluges and the creation of new lakes and seas across the continents. One of

Figure 3.4 Clay tablet with the Sumerian Flood story, ca. 1740 BC[39].

these theories, referred to by its discoverers as the Ryan-Pitman Theory,[40] claims that a catastrophic deluge about 5600 BC spilled water from the rising Mediterranean into the Black Sea, thereby significantly expanding its shoreline to the north and west. Obviously such dramatic events had great social impact.

Another water body, the Red Sea, is quite significant in Jewish and also in Christian history. In the book Exodus it is told how God empowers Moses to part the sea, thereby enabling the Israelites to escape out of Egypt by crossing the Red Sea and escaping from the Egyptian army that was chasing them. While the Israelites could walk safely to the other side to their promised land, the Egyptians drowned as the sea came together again. This miracle illustrates the power that God has over nature, including the mighty oceans. Water in itself is considered powerful and is used here as an instrument of God.

3.5.4 Environmental view

Already two thousand years ago, before the environment became a worldwide human concern, Judaism addressed environmental issues. Talmud literature over the generations set regulations to combat air, water and noise pollution and to regulate waste disposal.[41] These mainly elaborated on the basic Jewish law laid down in the Torah that nobody has the right to cause unnecessary harm or discomfort to his neighbour's person or property. As such, there were regulations for air pollution to prevent harm caused by smoke drifting from one person's property into his neighbour's. As to water law, the Talmud includes a few references. Water is considered a divine gift. As it brings benefits to all living creatures, water resources are basically excluded from private ownership and belong to everyone. There are also restrictions prohibiting riparian landowners from planting anything at less than a certain minimum distance along an irrigation canal or along the banks of a navigable waterway. This was intended to both protect the waterworks and facilitate transport, as in those early days barges were pulled by haulers walking along the waterways. Regulations were also in place for the maintenance of water works, wells and irrigation ditches. Inhabitants of riparian areas had to assist each other, in relation to the benefits received. In case of maintenance works, downstream residents had to help those upstream with repairs, and in cases of drainage the roles were reversed. Orders of priorities were also established for the domestic use of water from a public town well, thereby giving drinking by humans precedence over watering the animals, which subsequently preceded the use of water for laundering purposes. These legal principles are very similar to those later developed under Islamic law. This is not surprising, considering the fact that laws reflect the needs of a society living in a certain context, which in this case concerned groups in the same region with similar climatic circumstances.

In general, the Talmud perspective on the environment states that while we may use the world for our needs, we may never irresponsibly damage or destroy the environment. A Jewish professor from the University of Calgary, Eliezer Segal, gave the following answer to a question posed to him by one of his students, who was concerned about a passage in the Hebrew Bible that man is entitled to exploit the earth:

> *Central to any assessment is the recognition that, however wise and relevant our ancient sources are, at times they reflect a reality that is fundamentally*

> *different from our own. Here, too, we should take care not to lose our historical perspective...*
>
> *Our forefathers were an agricultural folk. So if we do accept as a fact that Judaism is consistent in placing human interests above the natural world and in urging the exploitation of nature for human convenience, this should by no means necessitate a negligent attitude towards either the environment or natural resources.* **The desire to keep the world clean and fruitful is justified by the most selfish of interests: you cannot exploit what is no longer around.** *More importantly, even if one should have wished to ruin the ecological balance, pre-industrial technology simply did not have the means to produce such destruction. Until the present century not even the most perverted of intentions would have succeeded in destroying the ozone layer, saturating our food with harmful chemicals or polluting the Alaskan coastline. The kinds of issues that we associate with environmentalist policies were quite unimaginable two hundred years ago.*[42]

In this perspective Jewish tradition seems quite aware of the dependence on our natural environment. The wisdom of putting the holy words on environmental issues in their historical perspective also applies to other world religions, such as Christianity and Islam. In present times many environmental initiatives which aim at more effective management of water resources are being taken up within Jewish communities worldwide. A noteworthy one is the production of *The Big Green Jewish Website*.[43] The website was designed to promote environmentalism through an engagement with biblical, rabbinic and contemporary Jewish sources.

3.6 WATER AND CHRISTIANITY

3.6.1 Origin

Christianity began with Jesus of Nazareth, who lived in the land of Israel and preached the laws of Moses and of the existing Hebrew Bible, with an emphasis on a merciful, loving God.[44] After Jesus' crucifixion one of his disciples, Paul, was particularly active and successful in spreading his message to other countries. At the time, the Greek language was the intellectual spoken language throughout the Middle East, and his followers were soon called Christians. The Greeks were also the principal converts to Christianity. By around 60 AD Christianity had spread west and north to many parts of the Roman Empire. With over 2 billion adherents, Christianity is the world's largest religion today.[45] It is the predominant religion in Europe, the Americas, Southern Africa, the Philippines and Oceania and is growing rapidly in China, South Korea and Africa. Water played an important role in spreading the word to other continents, as Christians crossed the great seas such as the Mediterranean and the Pacific and Atlantic Oceans in their evangelistic missions.

As Christian people live in many parts of the world, they have adapted themselves and developed many different traditions. In fact, the rapid expansion of Christianity is often attributed to the successful strategy of blending together existing pagan rites and Christian religious practices. The "pagan" Germanic Christmas tree is a well known example. As to water, over time pagan wells became holy wells, and churches were built upon or beside them. These holy wells were often celebrated with festivals and rites on old pagan holy days. At the same time efforts were made to eradicate the still existing animist religions, which attributed divinity to nature.

Because of the local adaptations made, there is no overarching Christian culture. This likewise applies to other world religions. Today Christianity knows five main denominations: Roman Catholic, Eastern Orthodox, Oriental Orthodox, Anglican and Protestant.

3.6.2 Role of water

There are many references to water in the Christian Bible, and it plays a central role in Christian religious life.

The Old Testament of the Bible is similar to the Hebrew Bible, and thus the importance of water for cleansing and purification mentioned in the previous section also applies to Christianity. Christian tradition also includes the stories of the Great Flood (Genesis) and the exodus of the Israelites from Egypt through the Red Sea, carrying on the idea that water has the power to destroy evil and enemies. Furthermore, water is said to have been made on the first day of God's creation of the world. It brings life, heals and can wash away people's sins.

Water in Christianity is primarily associated with baptism. The Catholic Church believes that through baptism the sins of a person are annulled.[46] Other Christian denominations view baptism rather as a public declaration of faith and a sign of welcome into the Christian Church. Baptism is said to have its origin in the baptism of Jesus by John the Baptist in the River Jordan. In those days baptism meant total immersion of the body under water, a practice still used today by the Orthodox churches. In most Western churches today one's head is simply sprinkled with water.

The use of holy water other than for baptism became a part of liturgical tradition in the 4th century AD. From the 9th century onwards, basins for holy water were placed at church entrances, from which people could take water to bless themselves

Figure 3.5 Jesus encounters a Samaritan woman at a well.

upon entering the church. In many churches blessed water is still used for sprinkling the congregation with holy water.

In the New Testament, the concept of "living water" is introduced, representing the spirit of God, being eternal life. This is symbolically described in the story from the Gospel of John, in which Jesus encounters a Samaritan woman at a well and offers her living water, the source of eternal life, so that she will never thirst again. In this story Jesus crosses over a number of cultural bridges: the holy, Jewish man reaches out to a sinful, Samaritan woman, thereby breaking down barriers of holiness, ethnicity, gender and religion.[47]

But what does Christianity today tell us about managing our water resources?

3.6.3 Environmental view

Environmental stewardship, or the responsibility to take good care of resources, including water, is an overarching and essential part of Christianity. The Bible says that God expects, even demands, Christians to be stewards of His creation. Stewardship can be measured by healthy ecosystems and sustainable, responsible consumption. For many decades water and other resources could be exploited without religious compunctions. There was a traditional view, shared by many Christians, in which God created nature for people to use for their own purposes. Recently, however, the tide has been changing, and religious thinkers have begun to reflect on how the broader values of their traditions might contribute to fostering greater care for the earth.

In September 2007 the leader of the Catholic Church, Pope Benedictus XVI, made a strong appeal to a large audience of young Catholics in Italy to save our planet before it is too late. He stated:

One of the fields in which it is urgent to work is most definitely that of safeguarding creation. Before it is too late, we must make courageous choices with a view to a strong alliance between man and the earth. This year, attention is directed towards **water**, *a most precious asset which, unless it is shared in a just and peaceful way, will become a cause for tensions and bitter conflicts.*[48]

This message gives a new impulse to the Church's responsibility to environmental stewardship and holds a promise for sustainable water management.

3.7 WATER AND ISLAM

3.7.1 Origin

It was also on the Arabian Peninsula that Islam originated in the seventh century AD. Although the Arabian Peninsula is surrounded by water (the Red Sea in the west, the Arabian Sea in the south and the Persian Gulf in the east) its overwhelming geographical feature is the lack of water. In fact, northern Arabia is one of the most inhospitable places on earth. With temperatures rising up to 50 degrees in summertime, frequent sandstorms and a shortage of water, it is a tough environment for any living creature. People well adapted to the harsh and demanding conditions are the nomadic Bedouin. For thousands of years the Bedouin have roamed the land, moving their herds from place to place in search of scarce resources and water.[49] Like most desert people, they

Interview:

Cultural and religious beliefs influencing water management in Pakistan

With Islam being the dominant religion in Pakistan, water is viewed as a basic need, with domestic use having first priority. As to rights, water is considered free for everybody, although people accept that you have to pay for the services delivered by the state. According to religious and cultural beliefs wasting water is not appreciated. In practice, the efficiency is not very high. This applies especially for a bulk water user, like irrigated agriculture. Recently farmers have been given some rights in managing the irrigation systems. There are now public water boards at canal level, in which farmers, government officials and other stakeholders are represented. They also have the legal right to collect water charges. What factors still undermine the effectiveness of water management in our country? I think in the first place, social issues like corruption, self-interest, and not to forget the very low level of education, with about 40% of our people being illiterate. A positive cultural factor is the collective response Pakistan people tend to have when droughts and floods occur, especially in rural areas.

Coming back to religion: We can acquire a lot of wisdom from religion to manage water resources in a way that benefits the people. I find the text in the Biblical book of Genesis, where Joseph interpreted Pharaoh's dream inspiring in this respect. I refer to the one where Joseph predicted seven years with plenty of rainfall and good harvest in Egypt followed by seven years with droughts and poor harvest and then he did wise planning and management to tackle the drought years.

Ilyas Masih, Pakistan

Figure 3.6 Farmers collectively removing sediment from a canal in Pakistan.

Figure 3.7 Mosques are centres for purification through prayer and water (Photo: Dubai, 2007).

have become experts in detecting the presence of underground water. As mentioned before, in the desert the secret of survival is to know where to find water.

The pre-Islamic Arabs were mostly polytheistic, worshipping e.g. the moon, the sun and the planet Venus as gods. Natural phenomena like caves, trees, waterholes and wells were also venerated. Judaism and Christianity were already known and practised in Arabia, mostly in the cities along the coast. Mecca was the centre of these religions. It was in this city in 610 AD that the prophet Mohammed, with the knowledge he had from the Hebrew Bible and the teachings of Christianity, began to preach the new faith of Islam. Soon an Islamic civilization spread across much of Asia, Africa, and parts of southeast Europe. Arab people devised a comprehensive approach to water management which had profound impacts in many parts of the world as Islamic religion spread.

3.7.2 Do's and don'ts

Today most Islamic countries are still situated in arid regions with the most water scarcity in the world. In this respect it is not surprising that the Islam knows a prohibition of the monopolization, spillage and pollution of water.[50] This is not only a matter of practical wisdom, water being the most precious commodity in the desert, but it is also considered a deed of devotion. In most Muslim societies, water is considered a "God-given" natural resource, one that should not be sold or bought. Hence, water should not be treated as an economic good but as a social good and cannot be

Water use is culturally determined

In Teheran there is no culture of efficient use. Conservation is not part of people's consciousness. Most people are religious; 98% is Muslim. The basis of Islamic ethics is conservation, but still people don't value this; people just use endlessly. Tariffs are a problem, as people don't want to pay much for their water. Until a few years ago water was ridiculously cheap. Today, the tariff for household use is still low. If you use more you pay a substantial amount more.

There is no problem with paying for water as such. People know they have to pay. The reason for the very high water consumption in Iran is that we want to get rid of our waste. In fact, we are very meticulous about being clean. Water re-use is not accepted. Some time ago there was a big scandal when people found out that waste water was used on crops. The biggest priority in our country has to do with the lack of education. We need to teach people ways to sustainably manage water. A start has been made with campaigns for efficient water use.

Assiyeh Tabatabai, Water supply and treatment engineer, Iran.

taxed. Yet although it is not possible to tax water itself, being a gift from God, it is perfectly legitimate to tax the water service or supply of it.[51] In Islamic countries water resources are public property, and a permit or concession is required for any use of water. As a result, all users are allowed to consume as much water as needed without paying for it.

There are, however, Islamic restrictions on the usage of water. The Koran gives clear directions about how people need to treat water.[50] Being a gift from God to mankind, this vital resource must be shared with others. It may not be monopolized or withheld from the needy or poor. The obligation to share water with others also includes providing water to animals. As in Christianity and Judaism, in Islam humankind has the first right to the resources that God has provided for His creation, followed in second place by cattle and household animals and third, the right to irrigation.[52]

At the same time, the spillage of water is condemned, as prophet Mohammed has said that a person should not waste water, even if one were to have the water of a complete river at his disposal. Furthermore, it is forbidden to pollute water in any way. In this respect the Koran states that God provides clear and clean water for humans and animals from which to drink and with which to purify themselves. Waste water re-use is allowed, provided that the water is treated to the extent that makes it safe for its intended use.[53]

3.7.3 Role of water

There are many more references to water in the Koran. The Arabic word for water, *Mâ*, alone is mentioned more than 60 times.[50] Water is considered the source from which all life originated, and its importance and benefits are highly praised. The teachings in the Koran have had far-reaching consequences for the Islamic way of life and its daily rituals. One of these rituals is the mandatory washing, before the prayers said five times daily, of those parts of the body generally exposed to dirt or dust (hand, feet, face, etc.). As dust is predominantly present in many Islamic countries, this is certainly not a superfluous routine. But water is not only meant to clean a person externally. It is also believed to clean spiritually and thus prepares a person for prayer. All mosques are equipped with large washrooms to perform this small washing or ablution, in Arabic referred to as *wûdu*.

The importance of water is clearly visible in Islamic art and architecture. Examples are the historic public baths in Islamic cities, the beautifully decorated public fountains with potable water and the renowned garden architecture with its lavish use of water. In contradiction to the Islamic prohibition on spillage of water, and given the scarcity of water, today water consumption levels in the Middle East, amounting to over 500 litres per person per day, are the highest in the world. Recently a Middle East newspaper stated that authorities and environmentalists worried about water scarcity in the Middle East have found a new ally – religion.[54] This relationship is in fact not quite new, as Islam has always advocated the prudent use of water. Plans were made to install water meters in mosques and government buildings in Qatar in order to keep track on water consumption – not for pricing purposes, but merely to raise awareness of water consumption.

3.7.4 Environmental view

Like Christianity, Islam is familiar with the concept of environmental stewardship. The Alliance of Religions and Conservation states on their website:[55]

Abundant water in Andalusia, Spain

In 711 AD Andalusia was invaded by Muslim Arab Berbers, popularly known as the Moors. They ruled the area for almost eight centuries, until 1492. Like most people used to living in the desert, they had a special bond with water. When the Moors set foot in Andalusia they thought they had entered paradise. Water was abundantly present, something they could only have dreamed of in the Moroccan desert. Water played an important role in their lives. They were convinced that economic prosperity and a high standard of living depended on a well-functioning water supply. To them, water should be omnipresent and available at all times: as drinking water, but also for their (Muslim) cleansing rituals before prayer. Water was also highly valued for its soothing nature. Unlike Western cultures, where people often find a leaking faucet irritating, the Arabic people in Andalusia found peace and comfort in the sparkling water of a fountain. Their culture was enriched by beautiful, well-irrigated gardens, by numerous fountains and public baths. They were masters in irrigation and were even able to divert rivers to the Arabic palaces like the famous Alhambra. Although much of this grandeur fell into decay over the years, its remains and influences are still visible in present-day Andalusia.

For the Muslim, humankind's role on earth is that of a Khalifah – vicegerent or trustee of Allah. We are Allah's stewards and agents on earth. We are not masters of this earth; it does not belong to us to do what we wish. It belongs to Allah, and He has entrusted us with its safekeeping.

From this statement it follows that Islamic people have a responsibility to conserve Allah's creation on earth. Nonetheless, the reality of water scarcity in most Middle Eastern countries today is concealed by religious, political and technological myths. In her book *Perceptions of Water in the Middle East*, Francesca de Châtel gives several reasons for this lack of urgency.[56] The absence of pricing policies in several Muslim countries, the continued political support for agriculture and water-thirsty crops such as wheat, rice and cotton, combined with the false sense of security created by large-scale engineering projects, all contribute to this mythology of plenty.

3.8 OTHER WORLD RELIGIONS

Apart from the main world religions described above, there are many more religions and non-religious beliefs that attribute specific holy or spiritual characteristics to water and have distinct views on its management. Two of these particularly worth mentioning are Taoism and the Baha'i faith.

3.8.1 Taoism

3.8.1.1 Origin

'*Be still like a mountain and flow like a great river*' is one of the statements laid down in the Tao Te Ching books by the presumed founder of Taoism, Lao Tse. For more than two thousand years Taoism, also referred to as Daoism, evolved in close interaction with – and was hence influenced by – other major Chinese religions, in particular Confucianism, Buddhism and Chinese folk religion. Taoism currently is primarily centred in Taiwan but has followers all over the world who have embraced its philosophical and religious concepts. The word Tao literally means "way" and as such Taoism represents a "way of life". At the same time, Tao is indefinable and unlimited. It is both the highest truth and the ultimate mystery. Like in other Eastern religions it unifies all seeming contradictions (yin and yang), which are believed to exist by the grace of each other and need to be in balance. In his book *The blood of the earth*, Dr. Allerd Stikker, founder and President of the Ecological Management Foundation, EMF, describes how this principle of Taoism inspired him and others with him to start a project in Taiwan to restore the balance between ecology and economy and later on to contribute to ecologically sound water solutions in China by setting up a Taoist Temple Alliance on Ecology Education.[57]

3.8.1.2 Environmental view

In Taoist philosophy, water appears as the essence of nature and a model for human conduct. Tao itself is often compared to water: clear, colourless, unremarkable, yet

Figure 3.8 Yin-yang symbol.

all beings depend on it for life, and even the hardest stone cannot stand in its way forever.[58] The ideal path to follow is the path which nature itself would follow. In this respect water is used as an ideal metaphor to describe the Tao principles, as water always follows its own course. It flows downhill and chooses the path of least resistance, flowing around obstacles and taking the shape of the vessel or container it enters. Like water, it is wise to "go with the flow" and not to try to struggle against the natural world.[59] In Taoism this is part of the "Wu-wei" concept, meaning "action without action" or "effortless doing". As to water management, Wu-wei also means not confining rivers too rigidly or diverting them from their course. This in contrast to Confucianism, which took a more dominating stance towards water and taught that rivers must be disciplined by constructing strong high dikes and forcing the water to move more rapidly toward the sea.

3.8.2 Baha'i

3.8.2.1 Origin

The Baha'i faith is the youngest of the world's religions. Having its origin in the Islam, the Baha'i faith first appeared at the end of the 19th century in Persia. It soon became an independent religion and spread to neighbouring Muslim countries, where Baha'i followers are still being prosecuted, and to northern India. Today there are almost six million Baha'i followers all over the world. Most of them live in India, the USA and Iran.

3.8.2.2 Environmental view

Baha'is view the world as one country with mankind as its citizens.[60] The strong sense of unity between all people (one global society) and the interconnectedness of all things are important pillars of the Baha'i faith. This is reflected in the Baha'i view on water management. It reads: [61]

The wise management of all the natural resources of the planet, including water, will require a global approach, since water is not a respecter of national boundaries. The use, sharing, protection and management of water need to be governed by spiritual principles of justice and equity and the fundamental concept of moderation. Decisions on water need to be taken through processes of consultation involving all those concerned or affected.

For Baha'is, respect for the creation in all its beauty and diversity is important, and water is a key element of that creation.[62] Water is essential to the functioning of all ecological communities, for agriculture and plays a key role in all the life support systems of our planet. As such, water is important for the preservation of the ecological balance of the world.

There are *Baha'i* laws concerning water and cleanliness through washing with water, laid down in the Bahá'u'lláh book of laws. These bear some resemblances with Islamic regulations, such as washing before daily prayers. Reference is also made in the book of laws to the water in public bath houses and household pools in 19th century Persia that was seldom changed. Using such water was strongly discouraged. One should use clean water that hasn't been used before.

According to Abdu'l-Baha:[62] *The Almighty Lord is the provider of water, and its maker, and hath decreed that it be used to quench man's thirst, but its use is dependent upon His will. If it should not be in conformity with His will, man is afflicted with a thirst which the oceans cannot quench.*

3.8.3 2012: The beginning of a new water era?

The Maya, an ancient civilization advanced in the use of writing, mathematics and astronomy, predicted already long ago that on 21 December 2012 the present Era will end. Their calendar knows a 26,000-year cycle, composed of five lesser cycles of 5,125 years each. The present and last cycle, running from 3133 BC to 2012 AD is called the Age of the Fifth Sun. On 21 December 2012 this cycle ends. This is also the day of the alignment of the December solstice sun with the Galactic equator and the beginning of the Age of Aquarius. Many people believe that at or around this date some catastrophic event will happen, such as a devastating flood that will engulf the world. Signs indicating to them that something is already happening are the recent increase in tectonic plate shifts, earthquakes, landslides, tsunamis, typhoons, cyclones, etc.[63] The number of floods and droughts has also increased dramatically in the last ten years. But no one can predict for sure if and what will happen in 2012.

3.9 CONCLUSIONS: FROM WATER WISDOM TO WISE WATER MANAGEMENT?

From the above sections we can conclude that water plays a key role in the beliefs and rituals of the world's main religions and spiritual beliefs. However, the question still remains:

How is it possible that in many societies with so much water wisdom this has not resulted in wise water management but rather the opposite? And can this trend be reversed?

Water, wealth and wisdom

Standing in a shady place in the desert of Qatar, Paul Buijs, whose company is supporting a test of different water qualities for crop irrigation, explains that desert people traditionally value water highly:

"Water used to be a scarce commodity and thus a precious one that shouldn't be wasted or polluted. This is in accordance with the religious and cultural notion that water is a gift, not to be spoiled. With the discovery and exploitation of the oil fields in the 1970s, the Middle East region quickly became wealthier. Along with the introduction of new water treatment technologies, such as desalination plants, a culture of plenty became the new standard. Within one generation the notion that water is a valuable commodity, not to be spilled, disappeared. At present the water consumption level in many Middle Eastern countries is the highest in the world. Public and private investments are directed at building more modern and bigger plants and water supply systems, rather than investing in the maintenance or repair of existing systems.

But bigger is not necessarily better. There is a risk of overdimensioning the capacity, as we have learned in Europe. It is also not a wise strategy to rely on just one strategy or water supply source, such as seawater. The challenge is to find an optimal combination of options, including both traditional and modern technologies as well as large-scale and small-scale solutions. And depending on need and availability, one should be able to switch between different water sources to guarantee a safe and reliable supply. There is surely a need for more sustainable options that fit in with the culture. As a technology exporting company, we also consider it our duty to point out water-saving options to our clients. Savings of 30–40% are readily feasible. These will not only minimise water consumption levels and waste water treatment volumes but also reduce the associated energy costs.

Raising awareness is a main challenge in creating a sustainable future of this region. Especially since not only the value of water but equally the price of water is low and short-term economic interests often tend to override long-term sustainable considerations. But there are signs of hope. At the government level there are directives to re-use waste water and not use water pure enough to drink for irrigation. Developers are compelled to build their own waste water recycling plants. Moreover, the same wealth that led to a general lack of interest in water saving practices will in due time also lead to a higher education level and increased awareness of water as a valuable resource, thus closing the cycle..."

For others planning to work on water projects in this region, Paul has the following suggestion:

"Knowing the local culture and context and investing in relations is essential for long-term success. As for my personal experience, there are more cultural resemblances than differences."

Paul Buijs, Global Membrains, working on a water irrigation project in Qatar.

Water is considered a purifier in most religions and used as such in ritual washings, bathing and for baptism. Clean and "living" water is recommended to be used for these purposes, but this is becoming scarcer than ever. Many religions also hold particular water bodies holy or sacred. Examples are the Ganges River in Hinduism or the Jordan River in Christianity. At the same time, these waters are becoming more and more polluted. The Islam knows a prohibition on spillage of water and advocates the prudent use of water, yet certain Islamic countries in the Middle East have the highest water consumption rates in the world. How can these paradoxes be explained and managed?

We may conclude that throughout the ages societies – especially in Western, "developed" countries, but also in urban cities in developing countries – have lost touch with nature. In Europe the "Golden Age" was the onset of unprecedented economic growth and exploitation of nature and the rest of the world. Urbanization and population growth, intensification of agriculture and industrialization, along with increasing food demands and mass consumption, have made the link between human needs and behaviour and the natural environment increasingly less visible. Much water wisdom was lost as a consequence. Moreover, in Western cultural belief, water is, generally speaking, viewed as an economic good or resource that can be exploited without moral compunctions.

3.9.1 Closing the circle

In today's world we can still find a harmonious relationship with nature in remote indigenous communities, living a traditional way of life, undisturbed by modern influences. They appear to have a deep respect for nature and the urge to protect water in particular.

At the same time, there is a tendency or movement, as one may call it, noticeable in many Western societies, towards a more natural and sustainable approach, at least for water management. Slowly, the conviction is taking root that large-scale technocratic solutions, such as hydropower dams, canals and other construction works, could prove to be disastrous in the long run. These projects may have large negative impacts on the livelihoods of local societies and the natural ecosystems in these regions, thereby disrupting the delicate balance between human societies and the surrounding natural environment. More and more, plans appear to give water more space and allow it to flow and meander freely. Waterways are also being made more natural and ecologically friendly to accommodate aquatic plant and animal species. In the 21st century we see a renewed interest in the values of some "old" traditional water practices and principles. We are thus closing the circle.

3.9.2 Religion: an alley for sustainable water management?

Religion can act as a promoter or inhibitor of new ideas. As to religious water wisdom: There is the realization that the wisdom laid down in the ancient holy scripts reflects a reality that is fundamentally different from the one of today and doesn't necessarily meet the needs of today's societies. Religions are now rethinking their roles, so as to make better contributions to the environment. And there are many keystones inherent to the religious beliefs themselves, one of them being the concept of environmental stewardship. Each faith has its own distinctive history and teachings, as well as its own unique relationship with the natural world. Yet most agree that freshwater resources are limited, vulnerable and important for life. The sustainable use of water depending on values such as fairness, equity and concern for others is in fact a recurring concept in most world religions. As mentioned above, religious leaders have begun to shine a light on how the broader values of their traditions might contribute to greater care for creation and the earth's resources. It's not just words; deeds are being done. This is illustrated by the many news headlines on this issue, for instance on the website of the Alliance of Religions and Conservation *(www.ARCworld.org)*. To mention just a few:

- *'First Islamic Conservation Guide launched – to help Muslim fishermen protect their seas'*
- *'"Eco-Sikh" leads Indian campaign to save the Kali Bein river'*
- *'Polluting the environment: one of 7 "new" mortal sins, Vatican says'*

Religion can thus become an alley to promote sustainable water practices, given the great reach and influence religion has on many of the world's communities. More than 80% of the world's population adheres to a religion. So if religious leaders took it upon them, as some are already doing, to promote water-saving practices, non-polluting policies, etc., this could make a real change. Many churches have become more sensitive in recent years to the sustainable use of the earth's resources, mainly from a viewpoint of stewardship of God's Creation. All over the world and drawing on their own traditions, religious communities are working in various ways to care for the environment.

The Alliance of Religions and Conservation, ARC, is supporting options for which temples, churches, mosques, etc. could become instrumental in improving access to water in their communities in rural and peri-urban areas.

So something is indeed happening, and the tide seems to be turning. Religion can no longer be used as an excuse to do nothing by claiming that mankind is entitled by God to exploit the earth or by proclaiming droughts or floods and their devastating effects as an "act of God", for which mankind is not to blame and can't do anything about. Because water is regarded as a "Gift of God" in many Islamic countries in the Middle East, the sense of urgency to use less water still seems to be lacking. Unless confronted with severe water stress, there is a general belief that God will provide water if needed. But this notion can also be used the other way round. The Director General of Water of Malaysia put it in a different context by saying:

> *The water that we receive from above is "a Gift of God", rather than an "Act of God" and therefore we shouldn't waste the water by letting it just disappear into the sea. If we retain the water that falls at times so abundantly in our country, it can be used in times of shortage. This helps to find a better balance between the wet season with too much water and the shortage of water during the dry season.*

From a cultural perspective it has always been a wise strategy to embrace traditional practices when trying to convince people to adopt new practices or beliefs. In fact, religion itself is the perfect example of how this works in practice as most world religions were able to spread their beliefs by doing so. So, if by example we want to convince Hindu people to keep the rivers clean, we need to find out for what rituals they use the water and if they would benefit from performing them with clean water.

What can we learn from indigenous religions' beliefs and their spiritual values attributed to water that could benefit the cause of sustainable water management worldwide?

Indigenous societies can teach other societies much about finding a development path that is environmentally sustainable. Having lived sustainably themselves for a long time, we can learn about the fundamentals of living in relationship with the natural world in ways that are responsibility-based rather than right-based and relation-based instead of consumption-based. It's not all harmony, though, as it is often argued that there's also a tendency within indigenous beliefs to exert control over the forces of nature.

Nonetheless, the importance and hence respect many indigenous people attach to water, and their practical knowledge on the relationship between humans and the natural world, may prove valuable for a new water ethic and sustainable management practice in the interest of human survival on this planet. On the other hand, for people being raised in a Western culture, the fundamental beliefs of indigenous people are

A Nueva Cultura del Aqua: A New Culture of Water in Spain

The semi-arid country of Spain has a tradition of rich historical water cultures. Since Neolithic times and throughout the Roman Empire and invasion of the Moors, water has played a major role in Spanish life. In a spiritual sense, as a valued necessity for agricultural irrigation, and for life and leisure purposes. At the beginning of the 20th century major changes took place. To overcome the economic and social crisis Spain was undergoing, irrigation projects were seen as the main solution to energize the nation's economy. These required major hydraulic public works (some 400 dams were built during the dictatorship of Franco) to meet the increasing water demand. Dams became the symbol of Spanish agricultural and industrial economic development. They represented the power of men over nature and the domination of rivers by means of technology. Yet the price paid was high. Many historical villages, magnificent canyons and beautiful valleys were lost and remained flooded forever. Many natural rivers and streams fell dry or became heavily polluted with pesticides and urban and industrial waste water. Zaragoza's Ebro River, once popular for bathing and fishing, became so filthy that in the 1980s no one dared to come in contact with its waters. The situation is slowly improving.

Since 2000 there have been massive demonstrations in Spanish cities for "A New Culture of Water in Spain". The widespread social upheaval was a reaction to the new National Hydrologic Plan (NHP) to construct more than 100 new reservoirs and divert large quantities of water from the Ebro River towards the Southern coastal provinces of Spain to meet the needs of intensive agriculture and coastal tourism. Protests culminated when in the summer of 2002 over 10,000 people from the Ebro Basin and other parts of Spain marched all the way to Brussels to try to stop the European funding of the Spanish NHP. They believed that it was in conflict with the European Water Framework Directive and demanded the application of "A New Culture of Water" in Spanish water policy. The peaceful protests of this social and scientific movement, referred to as the New Water Culture Foundation, were successful, and in 2004 the National Water Plan was withdrawn. The New Water Culture moves away from the perception of water as being simply a productive utility. It adopts a *holistic* view of water, including the broader ecological, cultural, spiritual and social functions of river systems. The social movement has clearly demonstrated the important role of culture as a trigger for transitional changes towards a more sustainable water management approach.

Source: Official Publication of the European Water Association (EWA) 2005 and article in Science Centric 17 September 2008: "A change in culture has spurred the process of water-related change in Spain"

relatively inaccessible, strange and hard to understand, as they are often far from their own reality. So although we may be able to borrow certain items via the wisdom of indigenous cultures, Western people may have to develop their own new set of values and beliefs to redirect their way of living into a long-term sustainable one. And of course, sometimes indigenous values will lose out to overriding financial pressures or other interests at stake, but they should at least be part of the equation.

Key messages

- Water plays a key role in most religious beliefs, values and rituals.
- Indigenous spiritual and religious beliefs hold a deep respect for water, and the traditional way of life often provides an example for a sustainable way of life.
- Understanding religious and non-religious beliefs regarding water is essential if we are to effectively deal with water management issues, e.g. the resistance to paying for water.
- Religious and spiritual wisdom can play a positive role in promoting wise water management by linking values and behaviour that promote sustainable water management.

4 – WATER: A SOURCE OF POWER

Chapter 4

Water: a source of power

All water flows into the ocean or into the purse of the rich.
Geographical origin Denmark

Swimming

Great plans are afoot:
A bridge will fly to span the North and South,
Turning a deep chasm into a thoroughfare;
To hold back Wushan's clouds and rain
Till a smooth lake rises in the narrow gorges.
The mountain goddess if she is still there
Will marvel at a world so changed.

Poem written in 1965 by China's former Communist Leader Mao Zedong after swimming the mighty Yangtze River.

4.1 INTRODUCTION

We all have experienced the natural power of water on occasion. It humbles us when we watch the force of a waterfall; it tumbles us when we stand in a swift current or fight the giant waves washing ashore. The intrinsic energy of running or falling water has been recognized since ancient times as a useful energy source. Controlling the earth's natural waters is also a fundamental human urge.

Throughout the ages and in many parts of the world control of the waters has even been used as a means to gain political power in the region. This chapter addresses water as a source of physical and political power alongside such issues as blue energy, hydropower dams and totalitarian water regimes. The key question is: What physical and political power can be attributed to water, what are the social implications and what does all this imply for sustainable water management?

In this chapter hydropower (4.2), dams (4.3) and political power (4.4) are discussed in relation to social and water management implications, followed by a concluding section (4.5).

4.2 HYDROPOWER

4.2.1 Falling water

Water's physical power becomes very clear when you experience a thundering waterfall like the famous Niagara Falls on the Canadian-American border or encounter the less well known but magnificent Takkakaw falls, the highest waterfall in the Canadian Rockies. Waterfalls are true wonders of nature, as our ancestors agreed. The word "Takkakaw" means 'It's a wonder' in the Cree Indian language and the name Niagara is said to mean 'Thunder of waters'.

The simple fact that gravity causes water to flow from higher elevations to lower elevations has led people to use running or falling water as an energy source. In fact, it's one of the oldest sources of energy. Centuries ago, people were already harnessing energy for purposes such as grinding grain in a mill by immersing a waterwheel in a stream. The steeper the fall and the faster the current, the more energy can be derived from it. Hydropower generation capacity is simply a function of the amount of flow and hydraulic head. This form of energy has been used and still is widely used all over the world, particularly in mountainous and hilly regions. Sometimes nature was helped by creating artificial water slopes, reservoirs and canals. Water power thus provided the local people with a relatively cheap and clean source of energy for activities like washing clothes, making paper, pulping sugar cane and mining.[64] By the beginning of the industrial revolution, around half a million water falls were powering Europe's factories and mines.[65]

Figure 4.1 Takkakaw falls, Canada.

4.2.2 Tidal water

But even in low-lying areas along the coast of France and England, as far back as the Middle Ages, people found ways to use the tidal movement of water to grind grain in so-called tidal mills.[66] They constructed dams in shallow creeks along the shore, and as the tide came in the gates swung open inwardly, allowing the sea water to fill the area behind the dam. As soon as the tide turned the gates swung shut, forcing the water through the tidal mill. The millers obviously had to be flexible, as their working hours were dictated by the time of the tides, shifting daily.

4.2.3 Pumping water

Water wheels were also ideal for pumping water from rivers for domestic purposes. In medieval times the arches of the famous London Bridge[65] housed water wheels that were used to pump water from the Thames River for the benefit of the citizens of London. Water power was also an important source of energy in ancient China. Large rotary mills, powered by water, appeared in China about the same time as in Europe, around the 2nd century BC.

4.2.4 Water power

In modern times hydroelectric power is generally produced by turbines that power generators to convert mechanical energy into electrical energy.[67] Back in 1961, the Niagara hydropower facility was the largest of its kind in the Western world, providing New York State's residents and local businesses with low-cost electricity. Today there are numerous hydropower dams and plants, with China being one of the leading countries in the world in applying this technology and exporting it to other countries. As a consequence of the imminent climate change, the energy crisis and the steep rise in oil prices, there is a renewed interest in hydropower. The latest ideas also include

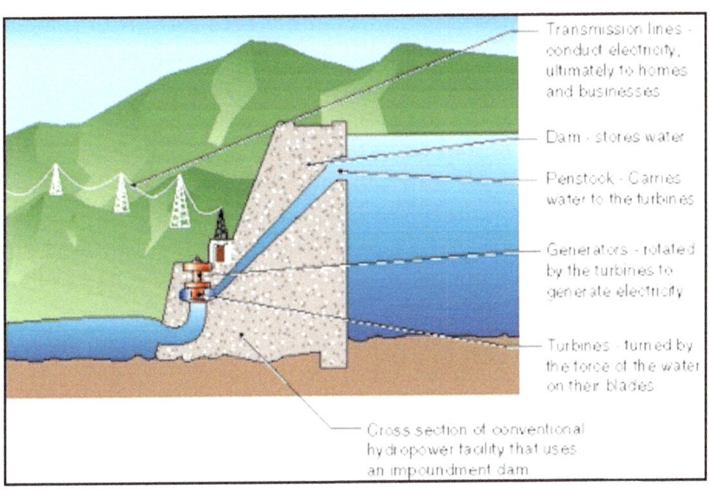

Figure 4.2 Cross section of a hydropower facility.

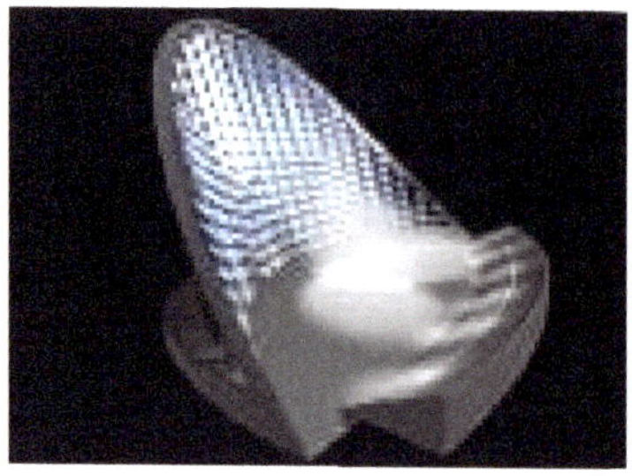

Figure 4.3 Artist's impression of the "Teatro del Agua".

plans in Norway and The Netherlands for using the difference in salinity at places where fresh river water meets salty sea water to create 'blue energy'. The process relies on osmosis through specific membranes, with brackish water as a by-product.

In this context the Teatro del Agua may also be mentioned.[68] This is an innovative and rather futuristic-looking desalination installation powered by sea, wind and sun. Driven by both sustainable energy considerations and the increasing shortage in potable drinking water in countries like Spain, the structure can be installed near the coast and produce fresh water on a municipal scale using renewable power.

4.3 DAMS

4.3.1 Brief history of dams[69]

The first evidence of dam building dates back eight thousand years ago. Old irrigation canals have been found in the foothills of the Zagros Mountains in former Mesopotamia. Most likely they were made by farmers in this area to control flows of irrigation water by using small weirs of brushwood and earth to divert water from streams into the canals. By 6,500 years ago, the Sumerians are believed to have installed a network of dams and irrigation canals in the plains along the lower Tigris and Euphrates. In Jordan remains have been found of dams that were supposedly built around 3000 BC to supply the town of Jawa with water. In the same region and time-period, around the time of the first great pyramids, the Egyptians constructed a water supply dam near Cairo, but this dam was washed away before construction was completed.

By the end of the first millennium BC, stone and earth dams had been built around the Mediterranean, in the Middle East, China, and Central America. The Romans were actually masters in building dams and aqueducts, the remains of which are still visible in countries like Spain, France and Italy.

South Asia, too, has a long history of dam building. There is evidence from the 4th century BC that long earthen embankments up to 34 metres in height were built to store water for Sri Lankan cities. King Parakrama Babu, a 12th century Singhalese ruler notorious as a tyrant and megalomaniac, boasted to have built and restored more than 4,000 dams in Sri Lanka. One old embankment he enlarged to a height of

15 metres and to an incredible length at that time of nearly 14 kilometres. No dam equalled it in volume until the early 20th century.

By this time dam building had become popular in other parts of the world as well. Almost 200 dams higher than 15 metres were built in fast–industrializing 19th century Britain. These were mainly constructed to store water for its expanding cities. Soon other countries in Europe followed. Up until the 20th century dam builders had little hydrological data and no computer modelling programs to rely on, and many dams, mostly earthen embankments, collapsed. Two hundred and fifty people were killed when a water supply dam in Yorkshire collapsed in 1864. The U.S. had a particularly bad track record, as one in ten embankments built before 1930 failed. More than 2,200 people were swept to their deaths when a dam above the town of Johnstown, Pennsylvania collapsed in 1889. The US was also the first to get a hydropower plant, a run-of-river dam in Appleton, Wisconsin, that began producing power in 1882. Over the next several decades many small hydropower dams were built along the swift-flowing rivers and streams in Europe, particularly in the Scandinavian country of Norway and in the Alps. Between the 1930s and 1970s two to three large dams opened every single day around the world. As dam engineering advanced, the size of the dams and power stations grew steadily. Around the world there are currently around 45,000 major dams in operation. The world's largest dam project, the Three Gorges Dam in China, is due to be completed in 2009.

4.3.2 World's largest dam: the Three Gorges Dam in China[70]

The first ideas to build a dam in the mighty Chinese Yangtze River had already emerged in 1919. A few decades later, Communist leader Mao Zedong was so excited about the idea of implementing this plan that he even wrote a poem on its behalf. As the communist and cultural revolution had exhausted all the country's financial

Figure 4.4 Artist's rendering of the Three Gorges Dam.[71]

resources, there was none left to realize the plans during his lifetime. It was therefore not until 1994 that the dam building project was actually begun.

The upstream portion of the Yangtze River with its famous gorges was selected as the perfect site to build the dam. For dam engineers the project presented the ultimate challenge. Once completed, the Three Gorges Dam will be the largest and most powerful hydro-electric power station in the world, thereby surpassing the Itaipu project in Brazil, currently the world's largest dam. The Chinese dam, made of concrete, measures over 2000 metres in length and 100 metres in height. When the water level will be at its maximum of 91 metres above river level by the end of 2008, the dam will create a water reservoir extending nearly 400 miles upstream and covering a surface area of just over 1000 km^2. It will provide electrical power to nine Chinese provinces and two cities, including the fast growing city of Shanghai. In addition, the dam is said to control flood events and to allow much larger freighters to navigate into China's heartland.

4.3.3 Environmental, cultural and social effects

But as usual, there is a downside, too. In order to build the Three Gorges Dam, over a million people had to be relocated. This obviously has a major impact on the social lives of these people and many farmers, being relocated to suburban centres, had to find new livelihoods. The dam is also causing the destruction of a valuable ecosystem. The habitat of endangered species like the Siberian Crane, that spends the winter in the surrounding wetlands, will be destroyed by the Three Gorges Dam. The dam has already contributed to the near-extinction of the Yangtze River dolphin. Another environmental issue associated with the dam is that of water pollution. The Yangtze is heavily polluted, as over one billion tonnes of waste water are being discharged annually into the river.[72] As the dam decreases the river's flushing capacity dramatically, the pollution levels will only increase. To overcome this problem and notably to reduce the water pollution from the densely populated city of Chongqing and its surrounding suburban area, more than 50 waste water treatment plants have been installed during the last few years. Sedimentation is another problem. Silt deposition in the reservoir will reduce the amount of silt transported by the Yangtze River to the Yangtze Delta which could result in erosion and sinking of coastal areas. The city of Shanghai, for instance, although 1,600 km away, rests on a bed of sediments and relies on further silt deposition to strengthen its foundation. In the near vicinity of the dam over a thousand archaeological sites will be submerged as the water level rises. Cultural and historical relics that are known and can be relocated are being moved to higher ground.

Then there is the risk of earthquakes and land slides, as the dam happens to be situated on a seismic fault. Lowering the water table followed by refilling the reservoir may induce seismic activity. This phenomenon has been observed previously, for instance at California's Oroville Dam, which was also constructed on an active fault line in the 1950s. When the reservoir's water supply was restored to full capacity after previous lowering for maintenance the area experienced a series of unusual earthquakes. At the Three Gorges Dam, 822 tremors were registered over a period of seven months after the September 2006 reservoir level increase. Luckily these were not severe enough to cause serious damage. The situation may worsen, though, by 2009, when the dam's water level is increased to full capacity and subsequently lowered during flood season. The increase in water pressure along with the water fluctuation may be just too much for the area to withstand.

4.3.4 A critical note on dams

> *'If the dam is constructed blocking the river, not only will the Salween River stop flowing, but so will Shan history. Our culture will disappear as our houses, temples, and farms are flooded.'*
>
> Shan refugee, Burma (2000)

The benefits of dams are quite clear. Dams allow people to create a reservoir for supplying water to households, agriculture and industries and to generate electricity cost-effectively. In particular, small dams have been found to be a sustainable and economic source of electricity, especially in rural areas. Other reasons for building dams include flood control, facilitating river navigation, creating reservoir fisheries and allowing recreational activities such as boating. Dams may even give a boost to the local economy.

Despite the obvious benefits, dams, and especially the large ones, are criticized for social and environmental reasons. The main reason for opposition worldwide is the forced eviction of huge numbers of people from their homes to make way for the huge water reservoirs created by dams.[73] An egregious example was the forced evacuation of local people by the Burmese army in 2007, using violence on a mass scale to clear the way for the construction of four large hydropower dams in the Salween River.[74] Locals were given only a few days' notice to leave, the sole message being: 'depart or die'. The operation formed part of a wider plan of Burma's military regime to gain full control over the exploitation of the nation's natural resources and to generate cheap electricity to be sold to its rich neighbour Thailand.[75]

> Over 3,500 people, including many ethnic Kayan, will be displaced by a new dam being built in the Pyinmana Hills that will give increased power to Burma's military leaders.
>
> Deployment of Burmese troops to provide security for the dam construction has caused increased forced labour and other abuses of local villagers, in direct contravention of the cease-fire agreement between the Kayan New Land Party (KNLP) and the Burmese military regime.
>
> The KNLP was originally formed in 1964 to protest another dam, Burma's first major hydropower project, the Mobye Dam in Karenni State, which flooded 114 villages, eventually driving many Kayan, including "long-necked" Padaung, to become refugees in Thailand.
>
> 'Forty years ago, we Kayan people lost our sacred lands to provide electricity to Rangoon. Now again the dwelling places of our guardian spirits will be submerged to power Naypyidaw', said Mu Kayan of the Kayan Women's Union.
>
> The Upper Paunglaung Dam, being built by the Yunnan Machinery and Export Co. Ltd (YMEC), is one of 24 major hydropower dams being planned or built by Chinese companies in Burma.
>
> Excerpt of a Press Release from Salween Watch
> June 18, 2008[75]

Economic considerations were also a primary reason for Burma's impoverished neighbour Laos to develop 30 hydropower plants, with plans for another 40 in the coming decades.[76] The electricity generated will similarly be sold to Thailand as well as to Vietnam, with their booming economies and resulting energy demands. In contrast to Burma, Laos endeavours to more carefully address the needs of humans and the environment. A good example is the Nam Theun 2 dam building project, supported by the national government and the World Bank. Evacuated people, mostly farmers, are being compensated and new villages have been built for them. Many animal species, such as elephants, apes, bears and leopards will be relocated to a nearby nature reserve, staying under the protection of the World Bank for the next 30 years.

Worldwide, dams are still the object of much criticism for changing the environment by affecting natural habitats. The huge water reservoirs inundate valuable land ecosystems, thereby destroying the habitat of the flora and fauna in the region. Furthermore, dams in a river may block the migration of fish and prevent fish like salmon from swimming upstream to their spawning grounds to reproduce. Currently fish traps are being placed in many dams to allow fish to migrate.

Millions of people around the world also suffer from the downstream impacts of dams, which can include loss of fisheries, decreased water quality and a decline in the fertility of farmlands and forests due to the loss of natural fertilizers and seasonal floods that healthy rivers provide.[77] In addition, critics doubt that the human and cultural losses are worth the projected benefits of many dams, arguing that both flood management and power generation can be achieved by faster, less expensive means.

4.3.5 Sustainable dam planning

The **World Commission on Dams** was the first independent international panel to assess the effectiveness of dams. They found that while 'dams have made an important and significant contribution to human development, and benefits derived from them have been considerable... in too many cases an unacceptable and often unnecessary price has been paid to secure those benefits, especially in social and environmental terms, by people displaced, by communities downstream, by taxpayers and by the natural environment.[6]

As a follow-up of this assessment, the World Commission on Dams developed internationally acceptable guidelines for the planning, construction and operation of dams. This entailed an approach to dam planning that would ensure affected communities are able to negotiate their own compensation packages and would be primary beneficiaries of dam projects' benefits.[78]

Some countries – for example, South Africa – have taken steps to incorporate the World Commission on Dams' recommendations into their planning processes for water and energy projects.

Where hydropower has proven to be cost-effective then it certainly should be one of the energy options to be considered, but only through a comprehensive and transparent planning process that takes into account the social, environmental and economic impacts of the project.

> In view of sustainable development it is important to assess the benefits and impacts of dam development on a river basin scale instead of on an individual project basis. The socio-economic and environmental aspects of upstream-downstream linkages should thereby be taken into account.

4.3.6 More power needed

There are many hydropower projects yet to come. Driven by economic considerations and the worldwide increasing need to use renewable energy sources, many countries have plans in stock to build new dams for hydropower generation. China is particularly active in this market and provides its engineering skills to other countries in exchange for resources like oil, minerals, copper and zinc.[79] These are needed to keep up with the rapid pace of economic development domestically. The African continent as a whole is desperate to gain access to electricity. There are now serious plans to build the world's biggest and most powerful hydroelectric dam and power plant ever in the Democratic Republic of Congo.[80] The Grand Inga project will be built on the Congo River alongside two existing hydroelectric plants and is expected to begin operating between 2020 and 2025. The plant in Congo could potentially generate twice the energy of China's Three Gorges dam and could help the estimated 500 million Africans without access to electricity. To make it a success the important social issues around the Grand Inga project must be dealt with satisfactorily. These include providing the poorer communities living closest to the source with electricity and compensating them for any negative repercussions to their livelihoods.

4.4 POLITICAL POWER

4.4.1 Crossing boundaries

Throughout the ages water has been used to exert political power and gain control over people and over entire regions. The Roman Empire thanked its expansion greatly to the ability to cross natural boundaries such as the Rhine River and the Danube by building bridges over them. Other engineering masterpieces at the time were the aqueducts – conduits on arches – that were built by the Roman rulers to channel water to the towns and cities. In ancient Rome's heyday, eleven aqueducts over 550 km long brought 190 million litres of water per day into the city that was further transported to the houses, fountains and baths, using lead pipes.[64] After the fall of the Roman Empire, enemy forces destroyed many of the aqueducts constructed by the Romans.

The ancient Greeks, Egyptians and Persians had already established the foundations for modern water resource management a few hundred years earlier. As to urban use, the Greeks were skilled in groundwater exploitation and knew how to effectively transport water, even from long distances to the cities, and treat it, before and after use.

4.4.2 Water as a natural defence

Water as a natural barrier is also used as a defence mechanism against hostile invasions. As such, rivers and marshlands have proven to be an effective barrier over the millennia. For instance in Prussia, the marshy lowlands, strengthened by fortresses built near the river, formed a natural defence against armies advancing from the east or west. The locals, knowing the territory, had a clear advantage in warfare over invading troops. In a similar way the Dutch Waterline, or *Nieuwe Hollandse Waterlinie,* formed a military line of defence of about 85 km that stretched from the Biesbosch in the South to the former Zuiderzee in the Northern part of Holland. Thanks to an ingenious water management system comprising locks, flood canals, existing waterways and dikes the waterline could be completely inundated within only three weeks.[81] Forty centimetres of water was enough to make the land treacherous and difficult to pass for soldiers, vehicles and horses. At the same time it was too shallow to navigate by ship. In addition the line included forts, fortified cities, bunkers and shelters for the local people to retreat to.

At times water also became an object of war. In the Second World War, for instance, the hydroelectric power station of Vemork in Norway became an Allied target. The Allies thought it essential to block German research into atomic fission by halting the production of heavy water at the Vemork power station.

4.4.3 Water in totalitarian regimes

In his book *Engineers of the soul*, author Frank Westerman describes how Soviet engineers in Stalin's Soviet Russia worked on plans to reverse the flow of the five main Russian rivers from north to south in order to supply water to the deserts in the southern Soviet republics.[82] In the name of socialism, the desert would flourish and Moscow would become a naval seaport. Stalin used the "engineers of the soul" – the country's most famous writers – to support his plans. Together with the actual engineers they were supposed to realize the dream of the Communist paradise by transforming the landscape with ambitious waterworks. The book addresses the intimate relationship between gigantic water works and totalitarian dictatorship, which is not only unique to Soviet Russia. It takes a whole army of forced labour, sometimes slaves, to realize these.

A notorious example is the building of the Danube-Black Sea channel in Romania in 1950s Communist Romania. Thousands of political prisoners were forced to live in labour camps at the site and work on its excavation.[83]

Many eastern European countries faced a transition from communism to capitalism. This often resulted in a shift in water ownership (privatization!) and had an impact on peoples' mentality regarding water management, as illustrated in the text below.

Interview:

Lost values...

We don't have major water problems in our country', Ruzica explains, 'apart from the occasional floods. After the Second World War, under Communist regime, major infrastructural works were installed, like roads, rails, dams, dikes, drainage canals and water supply systems. Many young people eagerly volunteered to do this work. They were proud to do so and it gave them the opportunity to be amongst peers and make friends. Older people are now somewhat nostalgic about the "good old days".

Recently I noticed that in many Serbian villages situated along the river, people are using river banks for solid waste disposal. It has to do with the fact that only few cities and industries have waste or waste water treatment plants, and that municipalities often do not provide containers for solid waste disposal. But it is also a matter of values. Before the war started, they would have felt ashamed as it wasn't morally accepted. But, since the war, so many values have been degraded or lost that dumping garbage in the water is not considered a big sin anymore.

Ruzica Jacimovic,
Serbia

4.4.4 Relation between water and power

The desire to exert control over the waters has shaped cultures. It's not so much the excess but rather the scarcity of water that drove people to build enormous engineering works to get access to water from distant places. Control over water also provided a means to gain power over others. Because of the sporadic occurrence and unpredictable nature of floods, flood protection seems to have had less impact on the structure of society and power. Irrigation in arid regions, on the other hand, requiring a constant effort, has led to new social orders and division of power.[84] Where water control in the ancient desert world led to a major re-shaping of the environment through the creation of grand hydraulic works such as artificial dams and elaborate canal networks, political power came to rest in the hands of an elite ruling class. These required centralized managerial bureaucracies to operate. In the most extreme form these "hydraulic societies" became despotic regimes, exerting absolute control over the lower classes. The famous scholar Karl Wittfogel claimed that the ancient hydraulic societies were the precursors of the modern socialist dictatorships as in Stalin's Soviet Russia and Mao's China. Donald Worster in his article on 'Rivers of Empire' makes a distinction between three broad types of water control that have occurred so far in history, each with its own techniques, patterns of social relationships and arrangement of power.[84]

4.4.5 Local subsistence communities

The first and simplest type of irrigation society is based on a local subsistence economy, where communities grow food for their own consumption. Here water control relies on temporary structures and small-scale permanent works that interfere only minimally with the natural flow of streams. The authority over water distribution and management remains completely within the local community, who also have the skill and expertise required to build and maintain their water system. Decisions about water rest in the hands of the family groups or clans living in the community. A still working example of such a form of local subsistence based irrigation society can be found in Bali, where rice-paddy farmers organized themselves long ago in so-called Subaks. Other surviving remnants of irrigation communities can be found in Valencia, Spain or in the Americas. A typical characteristic of such irrigation societies is that everyone contributes to the small-scale water works and the rivers are left largely to follow their natural course.

4.4.6 Agrarian hydraulic societies

But sometimes local management didn't suffice, as downstream villages had to establish control over those living upstream if they were to get any water at all. Further development of irrigation works, such as building canals and dams, interfered with the natural flow of the rivers, forcing the water to take new routes. This in turn demanded the loose network of individual villages in a particular river basin to transfer power to a central organization. In these so-called irrigated agricultural or hydraulic societies it was usually the state that provided water to the villages, demanding payment in the form of money, crops and physical labour. In Egypt, thousands of men had to toil under the broiling sun to accomplish major water supply and irrigation works. The Pharaohs had organized the entire water system into standardized, administrative subdivisions under direct control of the central government in Alexandria. Its organizational structure much resembled a pyramid, with the state rulers at the top. While water flowed to the peasants, wealth flowed to the rulers in the capital city. They derived their power from the technical and managerial control they exercised over the waters. They added to their domain by constructing new water works into new territory, using forced labour. The rulers managed in this way to create an empire. This happened in the four millennia before Christ in some of the great desert lands in the world such as China, India and the Middle East. Religion was often used to remind the citizens that they should respect and obey their superiors and be thankful for their benevolent control. The dominant attitude of the hydraulic civilizations is well illustrated by an inscription on the tomb of the fabled Assyrian ruler Queen Semiramis:

'I constrained the mighty river to flow according to my will and led its water to fertilize the lands that had before been barren and without inhabitants'.[84]

In contrast to Egypt where the Pharaohs had only a single river, the Nile, to manage, the emperors of China had to deal with many more river systems. And as much of China receives plenty of rainfall, the country's expertise lies mainly in flood control works and canals. The rise of Imperial China, commencing with the Han dynasty in the third century BC, is linked to some forty major water projects to control the Yellow River. At the time there was some diversity of opinion on water management between two important schools; the dominant Confucian school and the smaller Taoist school. The Taoists urged their rulers not to use force or impose their will on local people. In accordance with Wu-Wei and the idea of moving with the flow of a stream and not against it, they were also not keen on structures that too rigidly confined rivers or altered their flow. The Taoist engineer Chia Jang, who lived during the Han dynasty, put it this way:

'Those who are good at controlling water give it the best opportunity to flow away, those who are good at controlling the people give them plenty of chance to talk'.[84]

Confucianism advocated a more active and domineering stance toward water and people alike. Acting mainly according to this latter point of view, China's leaders transformed their country into a powerful, wealthy and hierarchically organized empire.

4.4.7 Modern hydraulic societies

The early, hydraulic, agrarian societies developed in the pre-industrial period have mostly been replaced by modern, technical ones since the 1800s. Capitalism and industrialization, resulting in fast exploitation of nature, reshaped the land in Europe, the United States and Australia. Water was no longer respected and venerated as a divine source as it had been in local subsistence communities, or treated as an ally in a quest for political power as in agrarian states. Water had become merely a commodity that could be bought and sold on the free market. In modern western countries democracy is the norm, with no single all-powerful state in control but rather a "controlled", often decentralized, bureaucracy or water authority with elected representatives.

Yet in many countries nowadays the link between water control and social power is as evident as ever. As mentioned in the last section, there has been a vast increase in gigantic hydropower dams and plants across the world. And there is an on-going struggle to exert power over water resources in countries where water is scarce. The Euphrates-Tigris region is such an example where shared water resources between countries cause international tensions. By extensively damming the rivers originating in its country, Turkey not only gets control over the water flowing through these major rivers but also gains more control over the downstream countries of Syria, Iran and Iraq. With water availability shrinking, particularly across the Middle East, Africa and Asia, there is an increased chance of water becoming a source of conflict, as will be elaborated on in the next chapter.

Corruption is killing

In Indonesia...

Everyone in Indonesia knows about **KKN.** It's the Indonesian acronym for Korupsi, Kolusi and Nepotism (corruption, collusion and nepotism). It plays a role in the major floods that regularly occur and affects sustainable water management in general. How?

Corruption: When contractors are awarded a project by the local government they usually have to pay the local officials for this favor. For instance, when the government has a budget of 1 billion ruppiahs to build a dike, a protective wall or a drainage canal, half of it is actually spent and the other half of the budget just 'disappears'. Not surprisingly, flood protection is often inadequate, leading to the loss of many lives.

Collusion: Collusion is an agreement, usually secretive, which occurs between two or more persons to deceive, mislead, or defraud others of their legal rights, or to obtain an objective forbidden by law, typically involving fraud or gaining an unfair advantage. It can involve 'wage fixing, kickbacks, or misrepresenting the independence of the relationship between the colluding parties. Collusion between government and the private sector aggravates the problems.

Nepotism: Giving favors and jobs to family and friends, who may not have the right skills needed for the job leads to inferior water infrastructure and poor maintenance.

Corruption is more or less accepted in Indonesia as everyone knows public salaries are low, but still higher than the income of the majority in this country. **Social jealousy** is another cultural phenomenon occurring to an increasing extent. It manifested itself in Jakarta during the latest floods. As a result of the gap between the rich and the poor and between protected and unprotected areas, people sabotaged the pumps and broke down protective walls of the richer communities out of an urge for retribution. If we have to suffer... they should suffer as well.

Source: Report of Author's site visit to Indonesia, February 2008

In Egypt...

In September 2008 a massive rock avalanche buried a slum on the outskirts of Cairo. The avalanche was caused by the absence of a sewage system and the wastewater of the households 'eating away' the mountain, thus contributing to the instability of the rock plateau. People told journalists that many homes were built illegally near the cliff edge after a bribe was paid to a city council engineer. Aboul-Ela Amin Mohammed, the head of the earthquake department in Egypt, said the entire plateau is in danger of further collapse. 'It is not the first time or the last time', he told The Associated Press. 'The area is full of densely packed informal housing with no central sewer system...When the sewage touches the fragile surface of the limestone it changes its consistency into a flour-like paste'.

Source: Press releases in September 2008

Where else?

Corruption links with culture: variations in level of corruption are due to variations in social norms and preferences that have been internalized by the inhabitants of the (studied) countries.

Transparency International's *Global Corruption Report 2008* demonstrates that corruption is a cause and catalyst for the water crisis. It is also an indirect cause of failure, lack of action and inadequate water governance. While corruption is most evident in developing countries, the phenomenon is not limited to low-income countries. Also in Europe, America and Australia corrupt practices are not uncommon. Corruption affects all aspects of the water sector, from water resources management to drinking water services, irrigation and hydropower. It therefore puts the lives and livelihoods of billions of people at risk.

Source: Transparency International, Global corruption report 2008, www.transparency.org/publications

4.5 CONCLUSIONS

The intricate links between water and power, ranging from hydropower to political power have been touched upon in this chapter. Such links become most evident in countries that score high on the power distance index scale of Hofstede.[85] Power in this context refers to the extent to which the less powerful members of organizations and institutions (like a family) accept and expect that power is distributed unequally. In itself a powerful regime is not detrimental to sustainable water management as long as its leaders are exerting their power in a wise manner with a long-term view for the sustainable use of water. Misuse of power, self-interest and corruption are fundamental causes and catalysts of the global water crisis.

At the same time, empowerment of local people through taking part in decision making around responsible water use has proven to be an effective strategy for sustainable water management. The socio-economic and ecological consequences of large-scale infrastructural works like dams and other major water works should always be incorporated into the decision-making process. Not just on the local level, but rather on a river basin level.

Key messages

- Water used as hydropower can be a sustainable source of energy.
- Dam building for hydropower and water supply can have a major cultural and ecological impact; entire cultures and ecosystems may disappear.
- Positive examples of hydropower plants and dam construction are usually smaller scale projects, with relocation and compensation schemes.
- The benefits and negative impacts of dam development should be assessed on a river basin scale. The socio-economic aspects of upstream-downstream linkages should thereby be taken into account and form part of the decision-making process.
- Throughout history water has been used as a means to exert control over others.
- Political power is not detrimental to sustainable water management as long as it is wisely used and not misused.
- Lack of transparency (corruption) aggravates problems.
- A long-term vision of those in power is a prerequisite for sustainable water management.

5 – WATER: A SOURCE OF COOPERATION OR OF CONFLICT?

Figure 5.1 A block of buildings is demolished in Chongquing, China, to clear land that will be submerged by the Three Gorges Dam. (Getty Images, 2007)

continues with a description of effective water dispute resolution methods and techniques. Conclusions are given in section 5.6. Section 5.7 of this chapter will provide concrete tips and suggestions for water professionals who want to foster water cooperation or mitigate water disputes. Intercultural communication issues will be highlighted here as well.

5.2 WATER RIGHTS

'When it rains collect the water.'
Burmese proverb

5.2.1 Who owns the water?

Who is the owner of water? Does the kind of water source define the type of ownership or not? Ownership of water is a relatively new phenomenon in human history. It is interesting to study the way our thinking on water ownership has changed over the centuries: in most cultures water is and always has been a *res omnium communes*: a resource which belongs to everybody and to all living creatures. Even during the Roman Empire, water was regarded as such. It is only during the past few centuries in Europe that water has become a resource that belongs to someone.

During the Dutch Golden Age, the perception that nature could be controlled was growing. During that time, water "ownership" developed from a *res omnium communes* to a *res publicae,* a resource which is available to citizens. Non-citizens had lesser rights, apparently also with regard to water supply. In the past century water has changed into a paid-for commodity, provided by a public, parastatal or private organization. In the past decennia privatization of water sources has become more common. The picture below shows an uprising of consumers in Cochabamba, the third largest city in Bolivia, protesting against the rise in water prices demanded by the commercial water company *Aguas del Tunari*, which was selected as the concessionaire to provide water services to this city in 1999. The development from a *natural resource for all* into *a resource for some* to a *resource for those who can pay for it* shows the "growth" of water as a commodity.

Figure 5.2 Battle on water dispute Cochabamba, Bolivia. (Photo: Tom Kruse)

Do you know which water stock - if you had invested only $1,000 in 1997- would be worth $40,760 today?!

A lady in Colombia selling water at the local market

Figure 5.3 Water has become a commodity.

At the outset of the 21st century, with our growing interest in traditional customs and acknowledgement of many traditional (knowledge) systems, we seem to have come full-circle: '*Access to safe drinking water for all*' is the slogan of the new millennium, which seems to imply that (clean!) water again is to become a *res omnium communes*. Nevertheless, the increased interest of investors and banks in water resources and water companies seems to indicate that privatization of water is a trend that will be difficult to change. In light of these developments, water has also been referred to as the "new gold" of the 21st Century.

Interview:
A Cartel of Drugs, Gasoline, Smuggling and … Water

'The city of Maicao, La Guajira in Colombia is quite special in terms of culture. It has Caribbean people, Venezuelans, Guayuu Tribes, Lebanese immigrants and, some years ago, we also had guerillas; all of them living close to each other – and all eager to protect their own interests.

About a decade ago a group of people took over the control over a water tank and started selling canned water from it to the local citizens, for about 1000 times the price we used to pay in the big cities. Working as a consultant for a water service company we were asked to help out. We revitalized the water supply system and soon started providing treated water to the local people through the piped system. The cartel thus lost their 'clients' and their profitable water business. And the people were happy that they received good quality water in their homes at a fair price.'

Leonardo Alfonso,
Colombia

5.2.1.1 Water rights

The ways water rights are defined differ according to the type of source and the culture of the people dependant on the source: whether it be wells, aquifers, rivers, lakes or seas. Rights may be held by individuals or by organizations. They may last for a limited time or endure in perpetuity, and may vary according to season and water availability. Rights may apply only to a specific use and parcel of land, or may be flexible in use and transferable.

Water quality may be specified or left out. Rights inherently bring duties and responsibilities.[86]

Wells

The rights people enjoy over water resources depend on the type of source or facility and the ecological situation. In the desert area of the Sahara everybody has access to a well, but pastoralists have their own, centuries-old trails that pass particular wells, while other groups use other wells. When no stress factors change the system, this works relatively well.

River rights

Creeks and river rights are quite common. Customary rights are known, and are often a matter of power. Mrs. Anabi from Nigeria indicated the following: '*Drinking water is fetched in the morning because then the pollution has gone. In the afternoon you can do your laundry or take your bath. Water distribution is according to power. All people tend to come at the same time at the place where it is easiest and safe to fetch the water. When you have to struggle to get your water the guys will help you*'.

The riparian principle is prevalent in many countries. English water law is based on the notion of the riparian principle: an owner of riparian land (land adjacent to a stream) has the right to use the water passing through his or her land. Roman-Dutch

Figure 5.2 Battle on water dispute Cochabamba, Bolivia. (Photo: Tom Kruse)

Do you know which water stock - if you had invested only $1,000 in 1997- would be worth $40,760 today?!

A lady in Colombia selling water at the local market

Figure 5.3 Water has become a commodity.

At the outset of the 21st century, with our growing interest in traditional customs and acknowledgement of many traditional (knowledge) systems, we seem to have come full-circle: '*Access to safe drinking water for all*' is the slogan of the new millennium, which seems to imply that (clean!) water again is to become a *res omnium communes*. Nevertheless, the increased interest of investors and banks in water resources and water companies seems to indicate that privatization of water is a trend that will be difficult to change. In light of these developments, water has also been referred to as the "new gold" of the 21st Century.

Interview:

A Cartel of Drugs, Gasoline, Smuggling and ... Water

'The city of Maicao, La Guajira in Colombia is quite special in terms of culture. It has Caribbean people, Venezuelans, Guayuu Tribes, Lebanese immigrants and, some years ago, we also had guerillas; all of them living close to each other – and all eager to protect their own interests.

About a decade ago a group of people took over the control over a water tank and started selling canned water from it to the local citizens, for about 1000 times the price we used to pay in the big cities. Working as a consultant for a water service company we were asked to help out. We revitalized the water supply system and soon started providing treated water to the local people through the piped system. The cartel thus lost their 'clients' and their profitable water business. And the people were happy that they received good quality water in their homes at a fair price.'

Leonardo Alfonso,
Colombia

5.2.1.1 Water rights

The ways water rights are defined differ according to the type of source and the culture of the people dependant on the source: whether it be wells, aquifers, rivers, lakes or seas. Rights may be held by individuals or by organizations. They may last for a limited time or endure in perpetuity, and may vary according to season and water availability. Rights may apply only to a specific use and parcel of land, or may be flexible in use and transferable.

Water quality may be specified or left out. Rights inherently bring duties and responsibilities.[86]

Wells

The rights people enjoy over water resources depend on the type of source or facility and the ecological situation. In the desert area of the Sahara everybody has access to a well, but pastoralists have their own, centuries-old trails that pass particular wells, while other groups use other wells. When no stress factors change the system, this works relatively well.

River rights

Creeks and river rights are quite common. Customary rights are known, and are often a matter of power. Mrs. Anabi from Nigeria indicated the following: '*Drinking water is fetched in the morning because then the pollution has gone. In the afternoon you can do your laundry or take your bath. Water distribution is according to power. All people tend to come at the same time at the place where it is easiest and safe to fetch the water. When you have to struggle to get your water the guys will help you*'.

The riparian principle is prevalent in many countries. English water law is based on the notion of the riparian principle: an owner of riparian land (land adjacent to a stream) has the right to use the water passing through his or her land. Roman-Dutch water law, on the other hand, is based on a grant of the right to use water, which is regarded as public property.[87]

Rivers which end in oceans get polluted either by oil spills or by urban or industrial activities. As Greaves stated, water rights are often cultural rights. To the Lummi, living in the Pacific Northwest south of the Canadian border, the salmon is part of their culture. This is also common to other first nation people near the west coast. Pollution of the rivers through large industrial and urban areas has a direct impact on the salmon population and thus on the culture of these and other societies.[88]

The Kwakiutl Indians, like many other northwest Canadian populations, rely on salmon for their economy. The salmon is, however, also very much linked to their cultural system, playing a prime role, for example, in the *potlatch* (a festival ceremony practiced by Indigenous peoples of the Pacific Northwest Coast in North America). As illustrated by Greaves (1998), the destruction of the salmon implies the destruction of their culture.

In order to streamline the use of large rivers and in order to prevent possible conflicts, many riparian countries have established international treaties. International treaties such as the Indus treaty, the Pak-Afghan Water Treaty, the Nile Basin Initiative, etc., frequently have the function of water authorities. They regulate the use of the river, including traffic on the river for commercial use, hydropower, etc. In the case of large river basins in Europe, the European Water Framework Directive which came into force some years ago makes river basin management obligatory.

Sea rights

The fish in the oceans become scarcer every day. Cot fish is highly priced nowadays. European and Japanese fishing fleets go to foreign waters to fish their concessions, sometimes with disastrous effects on man and ecosystem. Especially along the coast of West Africa, the situation seems to worsen every day.

Artisan fishery is in danger, as signalled, for instance, by fishermen from Cape Verde Islands. As international fishing companies empty their seas, the income of these and other traditional fishermen has been affected. Furthermore, the hauling nets scrape the bottom of the sea, thus destroying the ecosystem. The United Nations and other organizations such as Greenpeace have been campaigning against these fishing

Figure 5.4 Effects of a major oilspill

industries. On the other hand, the economies in these countries need the foreign currency which comes from granting these concessions.

A right to "use" the sea also refers to the exploitation of large oil and gas fields. International oil industries have large interests in these marine areas. Local and regional rights to fishing waters have to compete with national or international interest in these rich oil fields. The impact of spills on local fishing communities is often regarded as "collateral damage". However, in many of these cultures, the threat to the ecosystem also puts the culture in danger. The *Exxon Valdez* catastrophe in 1990 in Alaska had a major effect on the ecosystem.

5.2.2 COMPETITIVE RIGHTS AND INTERESTS

Water rights often compete with other rights and interests, including:

- Competition between indigenous water rights versus modern water rights. Some of the conflicts mentioned above already indicate these competitive rights. They continue to exist because the locals often have less access to information and lack equal access to justice.

Interview:

The influence of a regime

Because we had over 50 years of communism, water conservation is still a problem. It is not so easy for people to understand that others need water too. People use a lot of water and they are not careful in preventing to use too much water and spill the water. This is one of the consequences of some 50 years of communism during which you did not have to pay according to how much you consumed. Nowadays, you have to pay for your water consumption. This encountered resistance in the beginning, but by now this has been accepted, but it seems to be difficult to change people's behaviour on water consumption.

This is the general picture. However, in a city like Craiova where water has to come from the mountains, people are more water conscious as they are aware that the water reservoir needs to be (re)filled.

Romanian Waters National Administration, called *Apele Române,* manages the water resources in Romania and is a departmental agency of the Ministry of Environment and Sustainable Development. In order to facilitate proper water management, we have installed water boards. However, these water boards are not selected by the people, but they are part of the government and selected through parliament. These water boards work according to basins, but are rather centrally administered.

Romanian people distinguish different waters as Romania has many mineral waters (like Govora, Borsa, Borsec), which create quite a market for Romanian mineral water. Every region has its own type or taste and people know quite well which water is good to cure a particular disease: water for the liver, or the kidneys, etc. Drinking bottled water has become a habit.

With respect to religious customs, such as blessing of the house (*epitheny*) and baptising a child, water is used as well, but in insignificant quantities.

Ioana Popescu, Senior Lecturer in Hydroinformatics,
UNESCO-IHE Delft (Romanian by origin)

- The interest in economic growth versus access to safe drinking water
- The access to free waterways versus national sovereignity and borders.

In 2002 water was declared a human right by the former UN Secretary Kofi Annan. Water is a fundamental human need and therefore a basic human right. Human rights ensure the fundamental freedoms and dignity of individuals and communities, including civil, political, cultural, social, and economic rights. By making it a rights-based approach, due emphasis is put on the right to clean water, which will help to realize the goals set forth.[89]

This is a far-reaching statement, as access to water is often used as a war tactic: history has many examples of disputing parties who have poisoned valuable water sources, or impeded access to rivers and steams. By declaring water a human right, one weakens the potential for it to be used as an instrument of warfare or as a crime against humanity.

One interesting chapter of the UN Declaration focuses on water as a right to take part in cultural life:

5.2.2.1 Water rights are cultural rights

As water is used in many cultures for specific purposes, water rights are often also cultural rights, as acknowledged by the World Health Organization:

Water for cultural practices (right to take part in cultural life)

General Comment 15 on the right to water requires that access to traditional water sources be protected from unlawful encroachment and pollution. This applies particularly to the access of indigenous peoples to water resources on their ancestral lands, and also embraces the right to follow traditional cultural practices, such as performing religious ceremonies with water, for example the Hindu washing rites on the river Ganges in India. The right to water is violated if governments fail to take adequate steps to safeguard the cultural identity of various ethnic or religious groups. Examples include the destruction, expropriation or pollution of water-related cultural sites by state or non-state actors, or the offering by state authorities of land titles to individual members of indigenous peoples when these peoples traditionally take a collective approach to using property and attendant water resources, thereby threatening the cultural identity and existence of the entire group.[90]

5.3 WATER DISTRIBUTION

As explained in the previous chapters, according to Islam ground water and surface water are to be regarded as a gift of God, and no one should be deprived of quenching his thirst. Water should not be sold. On the other hand, water which is provided by man (through containers, pipes, etc.) will be paid for: users will pay for the treatment, storage and delivery of water. The commodity itself is not paid for. As explained in chapter 2, Islam is the only monotheistic religion which stipulates this.[91]

How does the distribution of water take place? Water can be distributed in many ways:

- by priority of use among different categories
- by time allocation
- by volumetric allocation
- by status

5.3.1 Distribution of water by use

Many cultures know a system of water distribution by use. In decreasing order: drinking water for humans is prioritized over water for animals, for agriculture and for industrial use. A distribution which changes this system may interfere with the social system: for instance, a large industrial company has received a concession for exploiting a source or for using the water for industrial purposes, while the local communities need the water for drinking or watering their animals. The new water distribution is not in line with the values they adhere to and resistance is likely to occur. In a similar way, social jealousy may occur when wealthy people build their homes next to poorer communities and deprive them of their water rights.

5.3.2 Distribution by time

In the non-western world, distribution by time is another way of distributing water among users. A comparative study revealed that among the Berbers of Morocco and the Bedouin in the Palestinian/Israeli region, the way in which water is distributed differs[92] Whereas the Bedouin distribute water per volume, the Berber of Morocco quantify water in units of time. Every group or family has the right to fetch water for a particular time period of the day or the week according to a set schedule.

5.3.3 Distribution by volume

In the Western world, water is distributed by volumetric quantities. Water meters are used in every house, and customers pay per quantity used. In a society where an alternative distribution system is used (such as the two described above), resistance to a different distribution system may be expected.

Meters are often sabotaged. This is also due to the price-service level relationship as several studies have indicated. People are willing to pay if a continuous, good service level can be guaranteed by the water company.

In many countries prepaid meters are used, in South Africa, for example, as a way to recover debts on outstanding arrears and in order to promote water conservation: consumers pay a flat rate and will be billed according to their consumption afterwards. This system stimulates water conservation, as consumers may notice their actual use. However, this can also result in people feeling discriminated against as the wealthier people are not metered and may "spend" water abundantly (the New Apartheid[93]).

5.3.4 Distribution of water by status

In other countries water may be distributed by status: those higher in status have first choice. On the other hand, those who have status often gained this status because of their wealth, which often has its origin in more favourable water access. Even though those who have first choice could take it all for themselves, they have the responsibility of sharing the water with other, less privileged users.[94] In view of sustainable water management and water conservation, a question arises as to which system has the best options.

5.3.5 Sustainable options for the future

The water distribution system in use will define water conservation techniques: some distribution systems favour water conservation (like metering), while others do not provide that stimulus.

Traditional distribution systems are not always the best option in view of water conservation, equity norms and sustainability. A change in behaviour may be required in view of these objectives. However, one should bear in mind that these customary distribution systems have been in place for centuries. An overly strict rupture of a common practice may not only encounter resistance but may also have unexpected consequences. Known examples include wells running dry because of irrigation practises. A change in (distribution) behaviour may have an impact on cultural values, traditions, and artefacts, i.e., regarding the role and position of elders, the use of artisan artefacts such as baskets and buckets.

On the other hand, cultural values may impede a change in behaviour as well: as we have already seen, certain religious values do not favour payment for water; beliefs and values about physical purity/body cleanliness may lead to excessive showering, and so forth.

5.4 WATER CONFLICTS

Water is a vital resource and as such it is both a source of conflict and an instrument of warfare in conflicts, as will be illustrated below. How conflicts are perceived and how they are settled is culturally defined.

5.4.1 Water as a source of conflict

Water disputes arise between:

- upstream and downstream users
- between rural and urban settlements
- between nomads and agriculturalists
- between traditional and modern societies
- between domestic and industrial consumers
- between countries on an international level.

Conflicts between *upstream and downstream* users are well known. Often these disputes arise at a local level, such as with small scale farmers downstream who are dependent on the upstream settlements for their water. As recalled elsewhere, in the Marib area in Yemen two types of water users had a water dispute after a change in water distribution, which was induced by the construction of a newly built dam. These are less common but do take place between riparians of the major rivers, such as the Nile, the Rhine, the Danube, the Mekong and the Euphrates and Tigris rivers.[95]

There is a dependency of *urban centres on rural ecological services*, including resources such as water. Urban communities often depend on the water supplies shared with surrounding agricultural lands. In the context of climate change, this may bring urban consumers and industries into conflict against rural food producers in competition for shrinking water resources.

The relationship between *pastoralists and agricultural communities* often is a mutually beneficial interaction: cattle are watered at communal water sources, and the cattle's manure fertilizes the land of the agriculturalists.[96] Climate changes and human activities have led to degradation of natural resources, while the demographic pressure continues to increase. As a result, a social restructuring takes place: farmers keep more cattle; some traditional pastoralists shift to sedentary farming practices and others keep moving around as nomads to look for water and pasture for their cattle. The pressure on land and water use often leads to conflicts.[97] And increasingly so, for when former pastoralists become sedentary the pressure on the already scarce resources increases even more, inciting conflicts around these resources, including water.

When *traditional societies* are confronted with *modern society* or "the state", the perception of property differs. Traditional societies regard resources such as land

Interview:

Dealing with cultural issues and water supply effectively

'The water supply coverage in Tanzania varies from less than 55% in rural areas to about 75% in urban areas. Main problems are the lack of funds for making large investments to replace old systems or install new ones and the many illegal connections. Through awareness campaigns we managed to deal with this latter problem in an effective way. The extra income from new paying clients enabled us to improve our services. Convincing people to pay for water was a challenge in itself as in the past people used to fetch the water themselves and got it for free. Traditional beliefs supported this culture of not paying for water. The majority of Tanzania mainland is Christian but we also have about 40% Muslims and 10% other religions like Hindu, Buddhism and traditional believers. Islamic people attach a high value to water and as such you can't enter into the Mosque before you have washed yourself. Muslims also manage water in a very efficient way due to its value. Still a fair amount of water is used for purification and other rituals in mosques, but also in churches. To accommodate all groups we have introduced a blocked tariff system for religious, domestic, institutions and commercial use. Water used for religious purposes in churches and mosques is subsidized and thus costs far less than water used for economic purposes. This system works quite well.'

Anthony Sanga,
Water Distribution Engineer,
Tanzania

and water as communal resources, while in "modern" states these become state or private property. With this shift in connotation, water has to be paid for, and resistance results.

Conflicts also arise when water resources management does not take into account the different categories of users such as *industrial* and *domestic consumers* and the (often) unwritten rights to use the water. When a change occurs in the order of users, conflicts may arise, as in the Cochabamba case mentioned earlier. As indicated above, societies define who is first in line to consume the water. An external factor which changes this order, as for instance in the Mekong area case, will lead to disputes.[98]

Strangely enough, water conflicts between riparian countries are far less common as will be elaborated in the next section.

In conclusion, one can state that water conflicts arise either because the water source itself is a source of conflict (too much, too little, too polluted) or because other resources (economic resources such as minerals, hydropower) are in demand, and the water source plays a vital role in getting these resources. However, water is never the single – and hardly ever the major – cause of conflict. Most of the time, concomitant factors are underlying the water conflict, such as historical factors (a history of war and former disputes), which may escalate the conflict.

5.4.2 Water as instrument of warfare

Article 54 of the Geneva Convention

Article 54 of the Geneva Convention states: 'It is prohibited to attack, destroy or render useless objects indispensable to the survival of the civilian population' and includes foodstuffs, livestock, drinking water supplies and irrigation works'.

Article 147 of the Geneva Convention

Article 147 of the Geneva Convention stipulates that to 'willingly cause [civilians] great suffering or serious injury to body or health' is a 'grave breach', which, according to Article 146, requires all High Contracting Parties to 'search for persons alleged to have committed or to have ordered to commit such grave breaches' and must 'bring such person regardless of their nationality before their own courts'.

Though water may be a *source* of conflict, it may also be used as an *instrument of warfare* as has occurred in the distant past as well as in recent years. Water is unfortunately becoming a target for *terrorism* as well, like in the case of the Warsak Dam which the Taliban threatened to blow up, the main water supply for Peshawar in the North West Frontier Province in Pakistan.

There are numerous examples from the past of how water resources were used during wars and disputes in order to defeat the enemy.[99] For example, during the 6th century AD, as the Roman Empire began to decline, the Goths besieged Rome and cut off almost all of the aqueducts leading into the city. In 537 AD this siege was successful. The only aqueduct that continued to function was that of the Aqua Virgo, which ran entirely underground.

Water has been poisoned to get enemies to surrender, as was the case with typhoid pollution of the Acre aqueduct in Palestina in 1948.[100] In the Darfur area in Western Sudan, water shortage seems to be both a cause of conflict and an instrument of warfare, as Mujahedeen militia deliberately deprived communities of this vital necessity, resulting in the starvation of humans and cattle.

During Allied bombing campaigns on Iraq, the country's eight multi-purpose dams were hit repeatedly. This has destroyed flood control, municipal and industrial water storage, irrigation and hydroelectric power. Four of seven major pumping stations were destroyed, as were 31 municipal water and sewerage facilities – 20 in Baghdad, resulting in sewage pouring into the Tigris. Water purification plants were incapacitated throughout Iraq.

The Dutch Water Line

The Dutch Water Line was a series of water based defences conceived by Maurice of Nassau and realized by his half-brother Frederick Henry.

Early in the Eighty Years' War of Independence against Spain the Dutch had realized that flooding low lying areas formed an excellent defence against enemy troops, as was demonstrated, for example, during the siege of Leiden, 1574. In the latter half of the war when the economic heartland of the Dutch Republic (i.e. the province of Holland) had been freed of Spanish troops, Maurice of Orange Nassau planned to protect it with a line of flooded land protected by fortresses that ran from the Zuiderzee (presently IJsselmeer) down to the river Waal.

After the final defeat of Napoleon in 1815 at the Battle of Waterloo, the United Kingdom of The Netherlands was formed. Soon after, King William I decided to modernize the Water Line. The Water Line was partly shifted east of Utrecht. In the next 100 years the main Dutch defence line would be the new Water Line which was further extended and modernized in the 19th century with forts containing round gun towers reminiscent of Martello towers. The line was mobilized but never attacked during the Franco-Prussian war in 1870 and WW I. At the advent of the Second World War most of the earth and brick fortifications in the Water Line were too vulnerable to modern artillery and bombs to withstand a protracted siege. To remedy this, a large number of pillboxes were added. However, the Dutch had decided to use a more eastern main defence line, the Grebbe line, and reserved a secondary role for the Water Line. When the Grebbe line was broken on May 13th, the field army was withdrawn to the Water Line. However, modern warfare could circumvent fixed defence lines (cf. the French Maginot line). While the Dutch army was fighting a fixed battle at the Grebbe line, German airborne troops had captured the southern approaches into the heart of "Fortress Holland" by surprise – the key points being the bridges at Moerdijk, Dordrecht and Rotterdam. When resistance did not cease, the Germans forced the Dutch into surrender by aerial bombing of Rotterdam, threatening the same for Utrecht and Amsterdam. Therefore, during the Battle of the Netherlands in May 1940 there was no fighting at the line itself.

After the Second World War, the Dutch government redesigned the idea of a water line to counter a possible Soviet invasion. This third version of the water line was erected more to the east, at the IJssel in Gelderland. In case of an invasion, the water of the Rhine and the Waal were set to divert into the IJssel, flooding the river and bordering lands. The plan was never tested and was dismantled by the Dutch government in 1963.

Source: Wikipedia

5.4.2.1 Conflict perception

An interesting question is how conflicts are perceived: what is seen as a conflict is not the same but may differ per culture. For instance, face-saving issues are more prevalent in Asian cultures than in Europe and even within Europe differences in attitude and perception may exist. People from southern European countries have slightly different perceptions of conflicts than people coming from northern countries, who view conflicts as something to be avoided. And in some cultures, it is not appropriate to be frank about emotions or about the reason behind a misunderstanding (as in Japan). Conflicts may therefore arise when water professionals/ consultants show insulting behaviour towards their counterparts in these countries and projects.

Misunderstanding and conflicts may be perceived as an individual case or as a collective dispute. In collectivistic countries conflicts may be collectivistic by nature, meaning they have an impact on the whole clan or family. Resolving such disputes demands the involvement of more than the (two) individual parties in the conflict. Disputes are a clan or community affair and need to be treated as such (see below).

Saving face

A European consultant was on a consultancy mission in Japan. Just before landing he saw that many parts of the country were flooded. Keen as he was to assist the client, he started to talk about the work as soon as he had arrived. The business card he had just received from his Japanese counterpart was quickly put in his jacket, while he asked about the major flooding problem. The Japanese reacted as if he had not understood. The consultant again stressed the fact that he was there to help the Japanese with their problems. The Japanese kept denying there was any problem. The consultant started to become irritated. He had been flying all those hours for nothing...?

A conflict was about to arise.

Nonetheless, most of the time the conflict deals with the way the water is *distributed.*

Ground water and wells

When it comes to ground water resources and wells, disputes arise about ownership, overexploitation, water tariffs, etc. Within traditional cultures disputes arise about the distribution system: among the Berber groups mentioned earlier, disputes arise about who takes which share of the day or week; among the Negev Bedouin disputes are about the quantity of water and its distributions.

In Darfur, Sudan, the civil war in this western province is linked to valuable resources such as water. A centuries-old co-existence of pastoralists and agriculturalists has been torn apart by the dispute, which has taken untold lives over the past decades. He who controls the water system controls the population. Access to water is a tactic of war as well in this area.

Dams and hydro-electrical power

Examples of disputes over dams are numerous, ranging from the Hoover dam near Mexico and the El Cajon dam in Honduras to the large Three Gorges Dam in China and others. In the case of China, not only did the dam cost billions of dollars, but the socio-cultural price was very high, as with many dams.[101]

Hydro-electric power dams will deprive downstream communities of their water resources as shown in the dispute between Egypt and Ethiopia regarding the Nile. As was shown in a study on Marib, Yemen, the new irrigation system was technically appropriate but socially a disaster: the new design had changed the former water distribution system. As a result, the once rich sheiks were facing poverty. From an egalitarian point of view, a more equally spread income distribution could have been a development goal. The problem was that nobody had taken the social-economic and cultural consequences into account, with violence and aggression the result.

Sea

International disputes arise on the use of, for instance, natural resources in Arctic areas (the North and South Poles). Regional and local disputes have been recorded from many countries, like the case of the Cape Verde fishermen as demonstrated above. Other cases are known such as in Benin where the coastal *Ramsar* area needed protection. Without the consent of all stakeholders, this goal was never to be achieved: local communities such as fishing communities, women who sell fish and play a vital economic role, the water supply company, the Ministry of Natural resources, the Ministry of Health, etc., all played a role in a sustainable solution.

The *Potlatch* of the Pacific Northwest Indians has already been mentioned as an example how a dispute on sea resources may easily develop, since people's identity and dignity are at stake. And by depriving these communities of their means of subsistence, they are deprived of their culture. Through withdrawal of their water rights their culture is at stake).

Surface water

Disputes on surface water are common as well. In the Netherlands, for instance, the Water Board Rijnland had a seven year-long dispute on water quality improvement in the Reeuwijk lake area. The issue focussed on the fact that different stakeholders such as local authorities, local fishing groups, restaurant owners, yacht owners, nature clubs, etc. all had different interests, and they were unable to settle their dispute themselves. As a result the water quality was deteriorating, and the different interest groups kept each other in a stranglehold. Through mediation techniques this dispute was resolved, whereby all stakeholders agreed to some concessions, some with results over the long term (testing through pilot programmes) and some with a short-term impact (environmental behaviour, biological measures). Professionals and communities alike contributed ideas to solve the problem.

In conclusion, the perception of conflict may differ: for some the conflict is a face-saving issue, a power issue, interest-based or rights-based. It may be perceived as a conflict between individuals or between entire groups or clans.

Depending on the perceived conflict, people have different *conflict management strategies*.

5.5 WATER COOPERATION

As research has shown, water conflicts arise mainly at the local, regional and sub-regional levels. When major rivers, such as the Nile, the Rhine, the Danube and the Mekong are considered, the opposite seems to be true: these can be considered peace builders, or sources of cooperation.[102] In Europe the Convention on the International Commission for the Protection of the Rhine against Pollution (Bern Convention) was signed by the Rhine bordering countries as a basis of international law for future co-operation. These countries were Switzerland, France, Luxembourg, Germany and the Netherlands.

International Commission for the Protection of the Rhine (ICPR)

Nobody is more aware of the fact that water protection is an international affair than the Dutch. The Rhine pollution has always shown its particularly negative effects in the Netherlands. Fifty years ago the Dutch were already complaining about the contents of phenol and salt in the Rhine water which made supply of drinking water of large areas very difficult. That is why the Netherlands united the Rhine-bordering countries to a common meeting at an early point in order to discuss problems of water protection and to look for common solutions. The "International Commission for the Protection of the Rhine against Pollution" (ICPR) was founded in Basel on July 11, 1950.

ICPR

On an international level cooperation exists through water treaties which regulate economic activities on rivers and seas. This is institutionalized through treaties such as the Nile River Based Initiative, the Danube, the Mekong Commission, the Indus treaty, the Pak-Afghan Water Treaty, etc.[103] Another example is ECOWAS, aimed at trans-boundary water management in African countries.[104]

Since the adoption of the EU Framework Directive all countries of the European Union are using a river basin approach for water management. This requires co-operation within and between all upstream and downstream users and countries within a river basin. A good example is the Danube river basin, where International Commission for the Protection of the Danube River (ICPDR) acts as the co-ordination body for the development of a comprehensive management plan for the entire Danube river basin. This process involves experts from industry and agriculture, and representatives from environmental and consumer organizations as well as the local and national authorities. All river basin stakeholders have been invited to actively participate to shape the future of the Danube and its tributaries – for their common heritage.

Another interesting example is that of the Nile Basin Initiative, which brings together the 10 riparians of the Nile River basin for high-level water negotiations, an example of environmental cooperation that has been successful in preventing conflict over water in a region that is troubled by ethnic, political, and territorial conflicts.[105]

Many countries who share borders fall within the same river basin. As stated by Wolf, territory in 145 nations falls within international basins, and 33 countries are located almost entirely within these basins.[106] Contrary to what one would expect, evidence proves this interdependence does not lead to war. Researchers at Oregon State University have investigated interaction (conflictive or cooperative) between two or more nations that was driven by water in the last half-century. They found that the rate of cooperation overwhelms the incidence of acute conflict. In the last 50 years, only 37 disputes involved violence, and 30 of those occurred between Israel and one of its neighbours. The Middle East, however, is a major source of water conflicts.

> Outside of the Middle East, researchers found only five violent events while 157 treaties were negotiated and signed. The total number of water-related events between nations also favours cooperation: the 1,228 cooperative events dwarf the 507 conflict-related events. Despite the fiery rhetoric of politicians aimed more often at their own constituencies than at the enemy - most actions taken over water are mild. Of all the events, 62 percent are verbal, and more than two-thirds of these were not official statements.
>
> Source: A.T. Wolf et al. 2005

Water seems to be a greater pathway to peace than conflict in the world's international river basins. International cooperation around water has a long and successful history; some of the world's most vociferous enemies have negotiated water agreements. The institutions they have created are resilient, even when relations are strained. The Mekong Committee, for example, established by Cambodia, Laos, Thailand, and Vietnam in 1957, continued to exchange data and information on the river basin throughout the Vietnam War.

On a smaller geographic scale as well, water is known to be a catalyst for cooperation. Regional water authorities (or Water Boards) boards in the Netherlands are an excellent example of how parties may cooperate in order to achieve their goal, i.e. safety against floods by building dikes and dams and ensuring sufficient good-quality water.[107]

Successful "export" of such Water Boards to other countries and cultures is not guaranteed. Success depends largely on the cultural and broader context.

- Is an equalitarian system acceptable to the respective culture?
- How are the hierarchical lines drawn within the water board?
- Are public participation and taxation accepted?
- Who may participate and who may not?
- How will a collectivist society be informed about the decision-making process within the board?
- Will the introduction of a new local or regional government layer with own tasks and responsibilities be accepted by the authorities currently in charge of water management?
- How are political, socio-economic (corruption) and legal aspects taken into account in order to make this board successful?

Interview:

The Dutch Polder Model: Water, culture and cooperation

The Dutch are well known for their ability to build dikes and polders and even reclaim land from the sea on which they actually live. There is also a cultural tradition in the Netherlands of consensus decision making, referred to as the 'Polder model'. It is often associated with the way the Dutch have organised themselves throughout the ages in a combined effort to fight the water in this low-lying country. More than half of our country is situated below sea-level, so cooperation was inevitable to survive. *The strongest man in a community is able to get water in areas where water is scarce, but the strongest man can not defend himself alone against floods.* And there were many floods, especially between 800 and 1250 AD when our forefathers started to exploit the peat marshes for agricultural purposes and use peat for fuel or salt supply. The resulting land subsidence increased the occurrence of floods. This necessitated land protection measures, such as the building of dikes, dunes, polders and drainage channels.

By 1200 AD landowning farmers had organised themselves and started sharing tasks and responsibilities to protect their polders from flooding. From 1300 onward, the Counts of Holland and in some places the Bishops took the responsibility upon themselves to appoint local officials (dike-reeves as President/chairman and councillors) to ensure proper control and execution of the tasks by so-called 'Water Boards'. Only full landowners and in later ages noblemen were eligible for these highly esteemed functions. And the system of decentralised communal cooperation worked well. Even Napoleon realised this when he occupied and ruled our country for a period of time after 1795. He decided not to interfere with our water management system, partly because it was far too complex to organise it centrally. In the last few centuries the Water Boards withstood the test of time, including the turbulent period of industrialization, when new technologies made it possible to create even more and deeper polders. Windmills were replaced by steam and later on fuel-driven and electric systems to pump the water out of the polders.

During the last decennia the scope of the tasks of the water boards evolved from mere water quantity to integral water resource management, including water quality management and water related fields of interest such as spatial planning, nature and recreation. At the same time not only the scope but also the scale of the water boards increased, with nowadays 26 Water Boards left, covering the entire country and its waters. They are thus the oldest still existing functional democracy in the Netherlands, with financial independence guaranteed by the legal right to raise local taxes. This is based on the principle that those who have an interest and thus benefit from the activities have to pay and also have a say. Every four years tax payers can elect their local representatives in the Board of Councillors. Co-operation and seeking consensus in decision-making are still the norm today. Water thus shaped our landscape and our culture.

Henk Tiesinga, Dijkgraaf (President)
Regional Water Authority Zuiderzeeland,
The Netherland

These and many other questions have to be addressed prior to the establishment of a water board.

From the principle that change is more successful when it is closest to the prevailing system, water professionals should (first) investigate what is familiar to a particular culture and how the new institutional system can be compatible with existing (or forgotten) structures.

Traditional irrigation technology fosters cooperation

In many African villages there are taboos related to use and protection of water from traditional wells and ponds. One such taboo is the use of cement and concrete in rivers in Zimbabwe. Cement and concrete is said to harm the water spirits. In the Eastern Highlands of Zimbabwe the local people (Manica) are not allowed to use concrete in diversion weirs that take out river water for their irrigation furrows. This taboo is interesting from a water management perspective. The diversion weirs in the rivers are simple structures made of local materials. As a result they leak and collapse during the rainy season. Yet, the leaking weirs ensure that the rivers do not dry up completely which is good for the environment and for the irrigators that take out water further down the river. The fact that the diversion weirs have to be rebuilt annually seems to foster cooperation. Manica irrigation technology and water management are internally consistent and mutually reinforce principles of equity and ecological integrity, which are globally recognized dimensions of integrated water resources management.

Summary taken from Workshop article March 2005 Water´s vulnerable value in Africa, Prof. Pieter van der Zaag, UNESCO-IHE

5.6 WATER DISPUTE RESOLUTION

As stated by Patricia Kamere-Mbote, a legal researcher and teacher based in Nairobi, water management is conflict management.[108] We would like to add to this that water management = conflict management = interest management. In case water disputes do arise, they need to be resolved, and the question of which dispute resolution method is best suited arises.[109]

In case a water dispute arises there will be a range of conflict management strategies available to solve the dispute. These are:

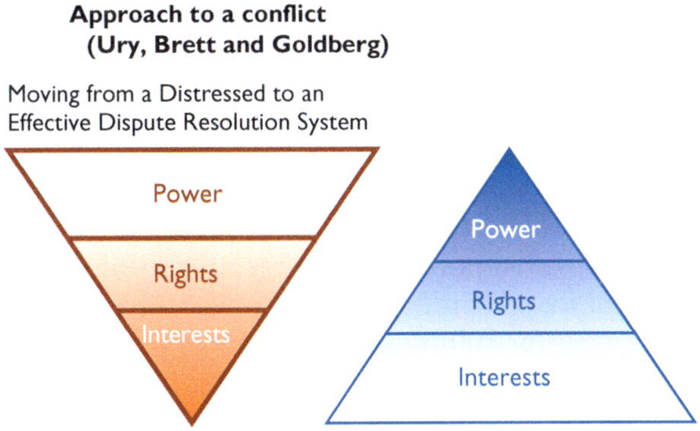

Figure 5.5 Three approaches to resolving disputes: interest, rights and power. (Ury, Brett and Goldberg, 1988)

- Use of **power:** for instance, by organizing strikes (Bolivia), going to the press, military enforcement (government, militias like in Darfur, etc.)
- Use of **rights:** for instance, filing a claim at court, litigation and other juridical procedures.
- Use of third party dispute resolution techniques by focusing on **interests,** like arbitration or mediation

Power based

The Pacific Institute for Studies in Development, Environment, and Security has investigated disputes over water which took place during the past 4500 years.[110] Less than 10% of these indicated conflicts were non-violent or not within a context of violences. So that (military) power was involved in most of the disputes. The study does not indicate how the conflicts were resolved, but it is most likely that power (force) was used.

Power is also used in a case where central authorities have political control over water sources. The role of a central authority is relatively recent, as in the past local communities had controlled their "own" water resources successfully for centuries. In countries where this shift from local to national control took place, the local people have to follow the central government in its decisions. Furthermore, in case of a dispute the indigenous people are in many cases in a less powerful position with respect to the authorities, who possess more knowledge and have better financial resources.

However, there are contrasting examples, such as the Lummi first nation people who have managed to successfully negotiate with the Canadian authorities for their right to use and own their water resources.[111]

Rights based

International water disputes are generally solved through water tribunals, international courts of justice, arbitration or through UN interventions. In this last case more mediation types of conflict resolution are probably used to mitigate the dispute. The oldest known water tribunal is that of Valencia which has been functioning for over 1000 years.

Local and regional water disputes are solved through legal procedures as well. Water boards play a vital role in preventing and settling disputes between farmers on water use.

Interest based

Distributive negotiation is often used to solve conflicts: both parties underline their arguments and facts in an attempt to convince the other. Unlike distributive negotiation, integrative negotiation according to the Harvard model consists of focussing on a maximization of possible solutions. According to this model, negotiation techniques such as mediation use integrative negotiation techniques instead of distributive negotiation which focuses more on positions. In many cases the conflict will be solved by focussing on positions, through the quality of the argumentation and through a verdict by the judge. Contrary to this approach conflicts can be solved through mediation techniques that focus on interests through questioning and interviewing. These conflicts may lead to reconciliation and forgiveness and are more forward-looking instead of looking to the past (who is the cause of the conflict, who is right/wrong).

In practice, a combination of these three methods of conflict resolution will be used: even when disputes are resolved through mediation or other interest based conflict management systems, the power and rights aspects are always in the background.

Diversity of conflict management systems

Many cultures employ a number of different conflict resolution systems. Depending on the type of conflict, a choice is made for a particular conflict resolution method. Especially when it comes to water disputes, many cultures have developed rather sophisticated, traditional methods, like the H'kim and shula in Islamic countries, the juju man in Sierra Leone (which also uses power in combination with mediation techniques), etc. Often, the respected head of the community is accepted as and has the status of conflict solver. Conflicts may also be solved by the courts in the capitals and major cities.

Traditional mediation is effective in dealing with interpersonal or inter-community conflicts. This approach has been used and is still being used at the grassroots level to settle disputes over land, water, grazing-land rights, fishing rights, marital problems, inheritance, ownership rights, murder, dowry, cattle raiding, theft, rape, banditry, and inter-ethnic and religious conflicts. As other dispute settlement methods have entered most societies, communities may choose which method to use. In Ghana for instance, depending on the type of conflict, individuals may choose for a traditional dispute resolution system or for the civil law (Anglo-Saxon) system.

Among the many "inventions" European countries took over from the Moors is the water tribunal (see also above). Driven by the desire for equitable distribution of water resources, effective social mechanisms for resolving disputes were developed in the Islamic world. Thus public water tribunals were established, which dealt with complaints of water users and passed judgement. They are still in existence today in the Andalusian city of Valencia.

Tips

1. Identify and utilize more experienced facilitators who are perceived as truly neutral. The World Bank's success in facilitating the Nile Basin Initiative suggests they have skills worth replicating in other basins.
2. Be willing to support a long process that might not produce quick or easily measurable results. Sweden's 20-year commitment to Africa's Great Lakes region is a model to emulate. Typical project cycles – often governed by shifting government administrations or political trends – are not long enough.
3. Ensure that the riparians themselves drive the process. Riparian nations require funders and facilitators who do not dominate the process and claim all the glory. Strengthening less powerful riparians' negotiating skills can help prevent disputes, as can strengthening the capacity of excluded, marginalized, or weaker groups to articulate their interests.
4. Strengthen water resource management. Capacity building – to generate and analyze data, develop sustainable water management plans, use conflict resolution techniques, or encourage stakeholder participation – should target water management institutions, local nongovernmental organizations, water users' associations, and religious groups.
5. Balance the benefits of closed-door, high-level negotiations with the benefits of including all stakeholders – NGOs, farmers, indigenous groups – throughout the process. Preventing severe conflicts requires informing or explicitly consulting all relevant stakeholders before making management decisions. Without such extensive and regular public participation, stakeholders might reject projects out of hand.
6. Water management is, by definition, conflict or rather interest management. For all the twenty-first century wizardry – dynamic modelling, remote sensing, geographic information systems, desalination, biotechnology, or demand management – and the new-found concern with globalization and privatization, the crux of water disputes is still little more than opening a diversion gate or garbage floating downstream. Obviously, there are no guarantees that the future will look like the past; water and conflict are undergoing slow but steady changes. An unprecedented number of people lack access to a safe, stable supply of water. Two to five million people die each year from water-related illness. Water use is shifting to less traditional sources such as deep fossil aquifers and wastewater reclamation. Conflict, too, is becoming less traditional, driven increasingly by internal or local pressures or, more subtly, by poverty and instability. These changes suggest that tomorrow's water disputes may look very different from today's.
7. No matter what the future holds, we do not need violent conflict to prove water is a matter of life and death. Water – being international, indispensable, and emotional – can serve as a cornerstone for confidence building and a potential entry point for peace. More research could help identify exactly how water best contributes to cooperation. With this, cooperative water resources management could be used more effectively to head off conflict and to support sustainable peace among nations.

Patricia Kameri-Mbote,
Founding Director of IELRC and
the Programme Director for Africa

5.6.1 Public participation

At a local or regional level, public participation is an approved method of preventing conflicts from arising.[112] Stakeholders discuss among themselves their needs and interests with (or without) the help of a facilitator and define a strategy for how to solve the problem and prevent a possible conflict of interests. In the coastal areas of Benin a multi-stakeholder process was started in order to find all stakeholders ready to work together to safeguard this Ramsar area prone to pollution, extensive fishing, etc.

In the Reeuwijk lake area in the Netherlands, a multi-stakeholder process was started to solve a dispute regarding the water quality. Local communities, environmental groups, the fishing association, the Water Board, and others were involved in this process.

Water Boards actually are a more institutionalized way of facilitating public participation, focussing on water management in a particular water catchment area. They have a similar objective: by involving the major stakeholders of a particular water system, the different interests can be dealt with effectively in the Board.

Interview:

Settling disputes in Iran

'Back in the old days people living in rural areas used to fetch the water themselves from a stream. Collecting water for domestic activities was a task for women and irrigation was and still is men's work. There was no management system, but distribution was perfectly planned based on a time basis. I remember that on a Wednesday water had to go to the village chief. On other days the rest of the community was given a certain time-slot. If you missed it, you missed it. Downstream people were usually poor and the richer people, who had more control over the water, lived upstream. My great grandfather was chief of the village. This meant that he also had to settle disputes, which sometimes occurred. The main tool he used for settling disputes was religion. All regulations are based on Islamic law. He particularly liked to use stories from the prophet's time and stories from the saints to settle any disputes.'

Assiyeh Tabatabai,
Water supply and treatment engineer,
Iran

5.7 CONCLUSIONS

From this chapter we may conclude that water may be a source of conflict as well as a source of cooperation. Water disputes are found mostly at the local, regional and sub-regional level, whereas water cooperation is more often found at an international level, leading to economic benefits for those involved. Conflict perception and conflict management strategies may differ from one culture to another. Water disputes can be solved through alternative dispute resolution methods, like [traditional] mediation.

The water professional who works in traditional rural areas should be conscious that traditional conflict resolution mechanisms have been in place for centuries and that such mechanisms should be acknowledged when a water dispute arises. For the professional who works in more modern and urban areas it is more likely that water disputes will be solved through western conflict resolution systems such as going to

court. Nonetheless, many water users do not have the financial means to go to court. In order to keep a balance between the two disputants, one should not too quickly turn to the court system, as this might weaken the position of the water users in the dispute

Tips for professionals

- Know the context and you will know what questions to ask.
- See if you can turn water into a source of cooperation rather than let it become a source of conflict.
- What is perceived as a conflict might not be the same as what western professionals will perceive as the conflict.
- Likewise, the solutions you offer may not be the solutions that will be accepted (and thus, these will not solve the conflict).
- Often local conflict resolution mechanisms are in place. Acknowledge these when necessary.
- Use *Intercultural communication techniques* (see chapter 6).
- Investigate cultural factors such as:

  - Who may decide? (power relation)
  - Collectivistic or individual issue?
  - Social relationships
  - Cultural values, norms and behaviour regarding water
  - Distributive or integrative negotiation

6 – WATER: A SOURCE OF SUSTAINABILITY

Chapter 6

Water: a source of sustainability

Te watra e kiri yu trutru, na watra wawan kan meki yu firi bun

'Real thirst can only be quenched by water', or: 'Each problem requires its own approach'

Surinam Odo[113]

6.1 INTRODUCTION

How can we make water management more sustainable by taking into account cultural factors? What factors should be considered? What can I do and what should others do? What are success factors for intercultural water management? These and other questions are addressed in this chapter.

In section 6.2 the concept of sustainable water management is recalled. Section 6.3 discusses the role of different actors in achieving sustainable water management. Section 6.4 gives practical guidelines on how to take cultural aspects into account and is followed by section 6.5, which mentions success factors for sustainable intercultural water management.

6.2 SUSTAINABLE WATER MANAGEMENT

In the first chapter the concept of sustainability was introduced. It has been defined as protecting and managing water as a valuable resource of the earth, for the use of present and future generations, while at the same time balancing the various interests and needs.[114] It was also stated that water has economic as well as ecological and social functions. As outlined in this book, water also has a strong cultural value in many societies. On the one hand, this cultural value of water (and sanitation) will be an enabling or countervailing power with respect to sustainable water management. On the other hand, water management itself may have a cultural impact as well, one which is sometimes positive but often negative. The challenge for the future is to find the right balance between all of these factors. This requires a combined effort by all actors and awareness of professionals of the interaction between these factors.

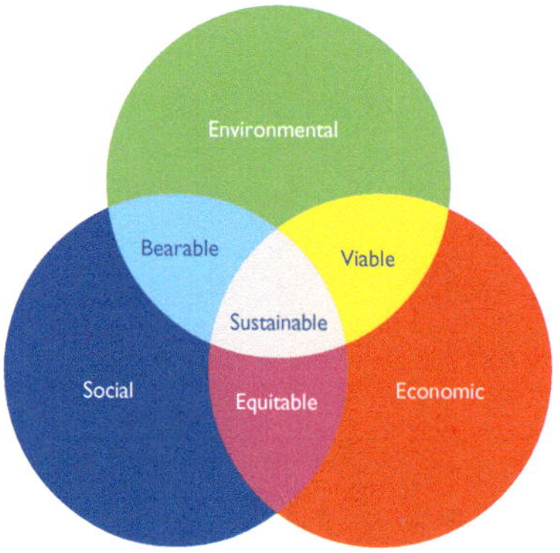

Figure 6.1 Sustainability is about finding a balance between social, environmental and financial considerations.

The Water Pyramid: a sustainable solution

The Water Pyramid makes use of simple technology to process clean drinking water out of salt, brackish or polluted water. Most of the energy needed to clean the water is obtained from the sun. The application is designed for tropical or subtropical regions around the world, where flat ground space is available, together with abundant radiation from the sun. Local entrepreneurs play a vital role in this integrated concept. Entrepreneurs distribute and sell the produced water, they have the knowledge of local habits and customs, and they understand the behaviour of the customers.

Source: Aqua-Aero Water Systems, The Netherlands

6.3 ROLES OF DIFFERENT ACTORS IN SUSTAINABLE WATER MANAGEMENT

The water problems in the world cannot be viewed in isolation or solved by engineers, scientists or policy makers alone. A combined effort is needed, as water is part of a complex web of political, economic and social issues. As a social issue it is aggravated by unbridled population growth, power misuse, corruption, pollution, poverty and lack of education. Only when different actors take up their respective responsibilities and start working together and in the same direction can things change for the better.

6.3.1 Role of governments

A stable, enabling environment is an important success factor for sustainable water management.[115] The reality is often far from this ideal scenario. The realm of realities ranges from governments run by capable and honest people to incompetent and corrupt officials. In some countries water management is centrally led, while in other countries this task has been delegated to decentralized local governments or water boards. Their authority may stretch from a small village to an entire river basin. Despite this diversity, the main role of any government or water management authority should be directed at enabling and facilitating sound and sustainable water management. This involves providing an adequate legal basis from national to local level and ensuring the necessary means and manpower to perform this task. It means investing in basic infrastructure for water protection, water treatment and supply. It also entails providing proper education to everyone and listening to the voice of the people and addressing their needs. These are all components of good governance, which in the context of water has been defined by the UN to include a participatory approach, transparency, equity and accountability, to mention just a few characteristics.[116] The most important thing is a long-term orientation and the political will to improve water resource management for the benefit of the people. WHO Director-General Dr. Margaret Chan stated in September 2008: '*More and more governments are determined to improve health by bringing water and sanitation to their poorest populations. Real improvements in access to safe drinking water have occurred in many of the countries of southern Africa*'.

6.3.2 Role of businesses

Businesses are often accused of acting out of mere self-interest by having a sole interest in profit-making and having no concern for people or planet. Although this may be the case in some instances, for example if a water supply company only provides services to richer communities that can afford to pay their bills, there are at least as many cases where private enterprises have shown that they can make a difference for the better. Companies are more successful when they are aware of the cultural values that impede or foster effective water use and sustainable water management.

More and more companies, especially the larger multinational ones, have embraced the concept of sustainability and responsible care. And many are dedicated to contributing to the achievement of the UN Millennium Goal targeted at access to clean drinking water. No matter what the drive, private businesses, usually have the network and resources to contribute to water management in an efficient and effective manner. Most companies are also experts in matching innovative solutions to community needs, thereby ensuring that local communities successfully adopt the new technology or system. This may range from flood protection technology to providing safe water supply and sanitation systems. At the local level, the micro-financing of local water management initiatives, particularly by women, has proven to be a golden match.

6.3.3 Role of foreign aid water professionals

Water professionals, such as skilled engineers and other water management experts involved in foreign aid programmes, have a special role to fulfil, and this is by no means an easy one. Dedicated to providing charitable or humanitarian aid to those being threatened by water or desperately in need of it, they may overlook some essential cultural aspects and contextual issues. Some tend to forget that what works perfectly well in their own country may not work at all and lie idle and unused in another country. This may be because there is a conflict with religious beliefs, or the system installed may disrupt the social hierarchy. It gets even more difficult if the local officials obstruct the plans, for instance if they are reluctant to transfer tasks and power to others. At other times it is simply a result of mistrust or a mismatch of partners working on a particular project. Careful assessment of the cultural factors before starting a project, and the involvement of local partners and inclusion of local knowledge from the beginning, may prevent huge failures from happening and makes aid programmes more effective. At the same time, water professionals should be competent in the field of intercultural communication. Development programmes should also focus on the socio-economic impacts, such as local employment generation, and carefully assess environmental and social impacts.

6.3.4 Role of scientists and teachers

Part of the problem with water in this world is a lack of knowledge, awareness and education. At present there are still large differences in the information level in different parts of the world. Internet may help to somewhat narrow the knowledge gaps. Science in itself can make a difference by developing the knowledge and know-how needed to deal with water management in a sustainable way. Education and public campaigns are vital to make people aware of the implications their actions or non-actions have on water safety, quantity, quality and their health. Water pollution is one of the major issues in this world where awareness-raising can lead to substantial improvements of water quality and diminish the associated health risks. Education programmes have shown that young children are an important target group, as they are still susceptible to learning and adopting new practices. Teachers at primary schools have a specific role in this respect and are familiar with what can be done and what is not acceptable from a cultural point of view. As natural change agents they can play a vital role in transferring effective messages to the target population. At the same time, teachers are valuable sources of information regarding the cultural values which enable or hinder sustainable water management. Foreign aid may also be directed at "train-the trainer" programmes for specific target groups.

6.3.5 Role of community

Every capable person in this world can contribute to help solve today's and tomorrow's water problems, either through his or her function or simply as a responsible citizen of this world. He or she can, for instance, contribute by not spilling or polluting water. Such responsible behaviour doesn't happen overnight and is often culturally determined. People are usually only susceptible to changing their behaviour when their basic needs of food, shelter and safety have been secured and they can see the benefit of it.

Public participation in water management boards or in new technical plans is crucial for local acceptance and long-term continuity. In Europe participatory water management has become a requirement for effective water resource use, planning and management. This requires the participation of all stakeholders, in particular the local community concerned, in decision-making processes. Participatory water management is gaining more support worldwide. Special attention must be given to the participation of women or other under-represented stakeholders. In most cultures men and women traditionally have different roles when it comes to water management and decision-making. Water programmes should define different scenarios with clear objectives (i.e. a sustainable water management project which targets the improved position of women as well, or water programs which focus on economic growth, etc.). Deliberate decision making on these set of objectives should be a prerequisite for every project. Furthermore, capacity building and training of employees and local villagers are essential for creating successful and sustainable projects. Technology implementation without proper training and engagement of local entrepreneurs and villagers easily leads to failure. Therefore new technology should be adapted to the local culture and situation.

6.3.6 Role of women and children

In traditional cultures women (and children) are the ones mostly responsible for water management for domestic use. As such, they select their water sources carefully on the basis of criteria such as economic considerations, perceived water quality and social relationships.[117] Knowledge of these criteria and an understanding of the selection process should be part of the design in cases where programmes aimed at water supply, treatment and sanitation want to see new services not just established, but also used, maintained and paid for. Although fetching water or using water for washing clothes is arduous, it provides women with an opportunity to meet and exchange information. Given the important social function of water for women, alternative means for communication may be needed when new water supply facilities at home limit women's meeting opportunities.[117]

The Hippo Roller used by women and children.

Sometimes we take for granted the way things are. Take, for instance, the long-standing 'tradition' of millions of women and girls in African countries to carry buckets of water on their head. They are forced to walk long distances on a daily basis to collect their water requirements for the day. But then suddenly someone somewhere on another continent invented the Hippo Water Roller. It is a barrel-shaped container with a clip-on steel handle designed to transport 90 litres of water. It's astounding in its simplicity. Approximately five times the normal amount of water can now be collected in less time with far less effort. It has already altered people's lives in many communities. Women are learning new skills in the time they used to spend collecting water. 'With the financial assistance of others, there's now the real possibility that a burden carried for so long can be set aside forever.'

Source: CNN coverage, www.hipporoller.org

6.3.7 Role of farmers and fishermen

With most of our fresh water worldwide being used for agriculture purposes, farmers are key players in achieving water resource sustainability. Sustainable water management practices include using water for crop production in a more efficient way and minimizing the use of pesticides and fertilizers polluting the water. Specialized drip systems and management practices (more crop per drop), water conservation practices and minimizing evaporation losses all contribute to a more efficient use of water. With the contribution of irrigation to world crop production, expected to increase in coming decades, farmers face a huge challenge. Water user associations in which farmers are represented can play an important role in preparing for sustainable water management and for responsible crop production and economic growth. Water policy reform may also be needed in order to allow more sustainable water management practices in agriculture.

Agriculture is the biggest water consumer. It uses approximately 70 percent of all fresh water withdrawals worldwide. The technical solutions to produce 'more crop per drop' exist. But often lacking are the investments and political will to improve rain-fed production, modernise irrigation systems and respond to the needs of people in rural areas.

Source: FAO/S. Maines/11491

Fishermen play vital roles in coastal areas. Industrial fish trailers destroy the ecosystem and leave the local fishing communities with reduced incomes. The traditional culture of these communities, where there is a delicate balance, is under pressure, for instance between the fishermen and the (female) fishmongers. Water management activities in the coastal areas of Northwest Canada are known for their impact on

Figure 6.2 Fisherman in Benin, Africa.

the culture of some of the indigenous communities who rely on fishing for their livelihood. The fishery sector can play a vital role in keeping sustainable ecosystems of seas and oceans, of rivers and lakes.

6.3.7.1 Working together

The above-mentioned stakeholders must work together at different levels to make water management more sustainable. For water professionals working on technical or institutional water systems abroad it is important to identify the significant players for each of the above-mentioned categories in the working area. They should be aware of the specific interests and culture of the key players and involve them from the start in the realization of any new plan. It is important to realize that although money is an important prerequisite to making things happen, when people take responsibility themselves, things will truly happen – and will last.

Ownership and capacity building

With respect to sustainable water management two items are of particular importance, namely: ownership and capacity building. Both at the individual user level as at institutional level a sense of ownership is relevant for success. If consumers do not feel a kind of ownership of a facility they will not feel responsible for its maintenance. On the other hand, if consumers do not know *how* to handle, take care and maintain the facilities, they will not be able to undertake these activities properly. Therefore, capacity building at the individual level through training, information, education and communication are important.

If it comes to the institutional level, a sense of ownership is relevant as well. When international consultants take the lead in the project, their counterparts will feel less responsibility for the water project.

Again, capacity building is a key factor to its success as well: without the knowledge and skills on proper project management, accountability, transparency and human resources management, a project is likely to fail. Therefore, both aspects should be an integrated part of any project.

6.4 HOW TO TAKE CULTURE INTO ACCOUNT IN WATER MANAGEMENT?

6.4.1 Understanding cultural differences

A useful tool for understanding differences in cultures across the world is the one developed by Prof. Geert Hofstede. He is one of the most cited authors regarding the world-wide classification of cultures. Hofstede defines culture as the '*collective mental programming of the mind which distinguishes one group or category of people from another*'. His classifications give insight in the average pattern of beliefs and values of a culture.

After a lengthy and extensive study at 50 national IBM offices in as many countries, Professor Hofstede found that cultures vary amongst five cultural dimensions, these being:

1. Individualism and collectivism;
2. Power-distance aspects;
3. Uncertainty avoidance;
4. Masculinity – femininity; and
5. Short term and long-term orientation.

There seems to be a correlation between the different dimensions, in such a way that collectivistic cultures tend to value large power differences and focus on uncertainty avoidance. Hofstede´s classification offers a valuable tool for recognizing and understanding the behaviour of groups and translating the information thus obtained to water management.

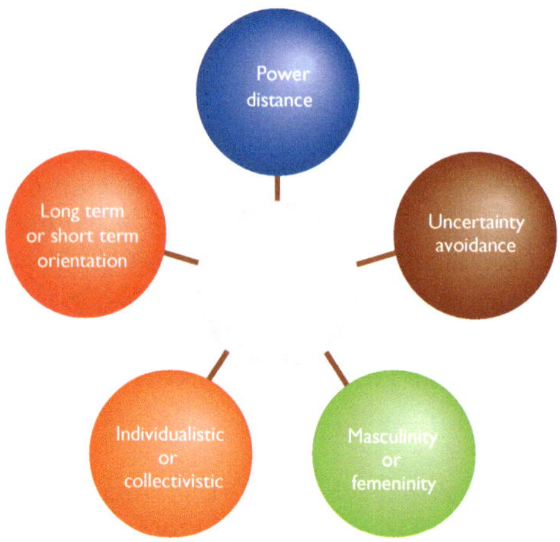

Figure 6.3 Hofstede's cultural dimensions.

Hofstede´s culture classification

Individualistic versus collectivistic

Hofstede distinguishes between individualistic cultures, which are more oriented towards satisfying individual needs and interests and, on the other side of the spectrum, societies that are more oriented towards collectivistic needs and interests. Individualist people tend to see themselves as independent. Personal preferences, needs, rights and goals are individualist's primary concerns; they place great value on personal freedom and achievement.

Collectivists, on the other hand, see themselves as interdependent; they place great value on maintaining group harmony. Norms, obligations and duties are main concerns. The interests of the individual are less important than those of the group.

Examples for water management: People in collectivistic societies have a collective and participative response to drought or flood problems; a collectivistic culture helps the (in-group) individual in, for instance, digging a well or cleaning a drainage canal.

And: The decision-making process surrounding water and sanitation projects in collectivistic societies requires the consent of all important group or clan members before a decision may be made.

'Not to wash one's dirty linen in public': Individuals within collective societies also try to avoid "losing face" for both themselves and their relatives.

Power distance

Power distance deals about the way hierarchical structures between groups or between individuals in a particular culture are accepted. The power distance index indicates the extent to which the less powerful members of organizations and institutions accept and expect that power is distributed unequally.

Example for water management: In a hierarchical society with a large power difference the principal of equity in water distribution, for instance, is usually not embraced.

Uncertainty avoidance

Hofstede defines uncertainty avoidance as the degree to which people feel threatened by the unknown or ambiguous situations and have developed beliefs, institutions, or rituals to avoid them. Avoidance-based cultures are very normative and have strong, formal rules and regulations, thus providing clear-cut guidelines to the people of that particular culture. On the philosophical and religious level there is a strong belief in absolute Truth; 'There can only be one truth and we have it'. The opposite type (cultures that accept uncertainty), are more tolerant of opinions different from what they are used to.

Example for water management: Introducing innovative water management approaches without paying due respect to the hierarchically highest in rank (social or formal) may cause one to lose face and may, encounter opposition. In Dutch there is a saying: 'What the farmer doesn't know, he doesn't eat'.

Feminine or masculine

Hofstede's next dimension focuses on the ways in which cultures manifest having either more masculine or more feminine characteristics. Masculine cultures promote inequality of roles between men and women. In cultures that are more masculine, values such as dominance, competition, ambition, progress, success, etc. are important. In more feminine cultures, more equality of the sexes is embraced. Important values are caretaking, solidarity, quality of life, etc.

Example for water management: The promotion of equal roles and equity between the sexes will encounter resistance in more masculine cultures.

Short-term or long-term orientation

This fifth dimension refers to the different time frames used by different people and organizations. It can be said to deal with Virtue regardless of Truth. Those with a short-term view are more inclined towards consumption and maintaining face by keeping up with the neighbours. With a long-term attitude, the focus is on preserving status-based relationships and thrift. Values associated with Long-Term Orientation are thrift and perseverance; values associated with Short-Term Orientation are respect for tradition, fulfilling social obligations and "saving face". Both the positively and the negatively rated values of this dimension are found in the teachings of Confucius, the most influential Chinese philosopher, who lived around 500 BC; however, the dimension also applies to countries without a Confucian heritage.

Example for water management: In societies with a short -term orientation, people will see to it that they get their share while maintaining face.

Cultural aspects based on G.J. Hofstede
http://www.geert-hofstede.com/

6.4.2 Applying cultural models to sustainable water management

Cultural values can be an enabling or a countervailing power with regard to sustainable water management. By taking these into account, water management programmes will be more successful in achieving its MDGs. But how should this be done?

Combining Hofstede's cultural factors with the contextual issues and other cultural factors, such as the core values and beliefs, traditions and practices of a society, results in figure 6.4:

How can this model be used for sustainable water management?

In general, a simple yet effective strategy towards sustainable water management practices is to consider the cultural dimensions and look at the:

1 Past: What cultural water practices have worked well within this country, and can we incorporate these in future water management plans?
2 Present: Talk to different stakeholders (including the local community) to find out what's really needed and what the counteracting and enabling factors are – both cultural and other (e.g. access to capital, legal support, core values and beliefs). Take these into account to ensure the acceptance of new water management plans.

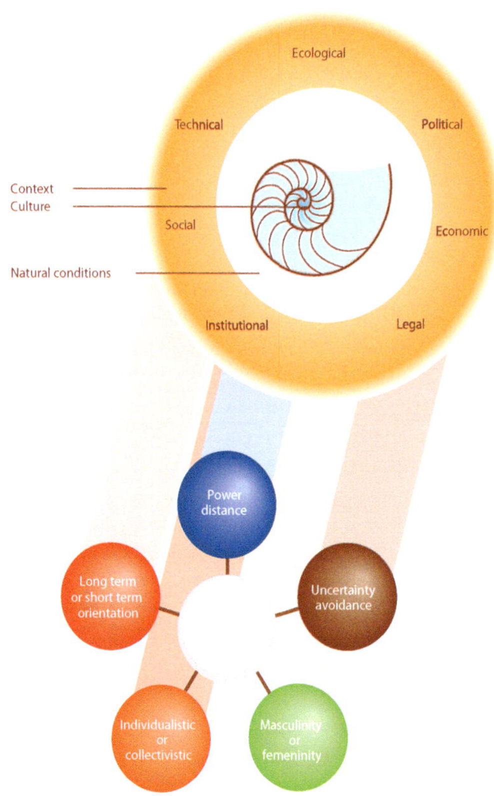

Figure 6.4 Cultural dimensions related to sustainable water management.

Example

Suppose a Dutch consultancy wants to establish a water board in a particular country, e.g. a Benin RAMSAR wetland area. The following is a minimum to consider:

- The national and international legal regulations regarding these sensitive ecological areas and the norms of the international community with regard to technical assistance (involvement of women in water and sanitation and other gender issues, participatory approaches, etc.);
- Socio-economical aspects: fisherman who rely for their livelihood on fishing local species, the women who earn an income by drying, smoking and selling fish, etc. But also: interrelationships between different groups, linkages, cultural aspects; and
- Institutional aspects: staff, capacities, working relationships, power relationships.

These, however, are not enough to guarantee the sustainable and participatory management of these resources. One should be aware of Hofstede's 5 classifications with regard to his actions. For instance, if there are plans involving a change in the water board or its members, one should be aware that the decision making is collectivistic by nature, that the power distance is an issue in this society, especially when it concerns in-group/ out-group boundaries (2 tribes). And thirdly, that the society as such does favour women becoming active members as long as this is not threatening to the other aspects (power, uncertainty avoidance, collectivistic focus, etc.).

This will imply that one has to discuss the issue with representatives of each group, investigate these issues and analyse whether they favour or disfavour the plan and which countervailing activities are needed.

3 Future: Use the outcome of the first two steps to realize the plans, thereby involving the local community as much as possible. Local acceptance, responsibility, ownership and involvement are an important key to success!

6.5 CULTURAL IMPACT ASSESSMENT

Cultural Impact Assessment is a useful technique in assessing – and subsequently mitigating – adverse cultural impacts of innovations and developments. For the purpose of this study, cultural impact assessment is defined as:

> *A process of identifying, predicting, evaluating and communicating the probable effects of a current or proposed development policy or action on the cultural life, institutions and resources of communities, then integrating the findings and conclusions into the planning and decision-making process, with a view to mitigating adverse impacts and enhancing positive outcomes.*[118]

This implies that decision makers should understand the cultural consequences of their decisions before they act and allow the people concerned to participate in designing their future. This ensures the sustainability of development actions and policies.

This approach requires a multidisciplinary effort and, according to Sagnia, involves the following in its better form:[118]

- Reviewing the existing state of the cultural environment and the characteristics of the proposed action and possible alternative actions;
- Predicting the state of the future cultural environment with and without the action (the difference between the two is the action's impact);
- Considering methods for avoiding, eliminating or reducing any adverse impacts and determining possible compensation for them;
- Preparing a cultural impact statement or cultural assessment report which discusses the issues and is used to inform and influence decision-making; and
- After a decision is made about whether or how the action should proceed, monitoring the impacts which do occur, and acting on the results of such monitoring.

What does this imply to both the engineer who is contracted to design and install a water system and to the professional who is contracted to guide the establishment of a water board?

Apart from the regular steps and actions necessary for good water management, the cultural context should be taken into account.[119] Also a Cultural Impact Assessment should be undertaken at the different project stages: the research, planning & design phase and the monitoring phase, such a Cultural Impact Assessment should be undertaken.

6.6 SUCCESSFUL COMMUNICATION REQUIRES CULTURAL COMPETENCE

> *I always believe we are the same; we are all human beings. Of course, there are many differences in cultural background or way of life, there may be differences in our faith, or we may be of a different colour, but we are human beings, consisting of the human body and the human mind. Our physical structure is the same, and our mind and our emotional nature are also the same. Wherever I meet people, I always have the feeling that I am encountering another human being, just like myself. I find it is much easier to communicate with others on that level. If we emphasize specific characteristics, like I am Tibetan or I am Buddhist, then there are differences. But those things are secondary. If we can leave the differences aside, I think we can easily communicate, exchange ideas and share experiences.*
>
> Dalai Lama

What do we need to know to prevent us from making huge mistakes in communication through ignorance of cultural factors?

It is often said that '*the more you understand of a particular culture, the more you realize what you do not know. It will be hardly impossible to get to know a different culture thoroughly*'.

Nevertheless, some people seem to be more intercultural competent than others and are able to communicate better with any people from a different culture. They possess constructive skills like empathy and good listening skills, combined with the ability to achieve a collaborative dialogue and to adapt quickly to new and unexpected situations. In fact, this corresponds with basic communication principles, such as treating people with respect, carefully listening to their concerns and needs and communicating on an equal footing.

But is it possible to enlarge one's own intercultural sensitivity? A good start, often used in cultural training sessions, is to get to know your own cultural background and values and determine how these influence your own behaviour and decisions. Using this as a reference point you will become more aware of the cultural values of other societies. You can gain some basic understanding about the other culture you (are about to) work with. There are many useful books and websites that one can look into in order to gain more knowledge about a particular culture prior to visiting a country. A noteworthy book series is *Culture shock!*,[120] which provide valuable and insightful information about many different countries and cultures around the world.

6.6.1 How to bridge cultural gaps and increase your own cultural sensitivity?

Bridging cultural gaps requires intercultural competences, such as the ability to recognize and acknowledge cultural differences. It also involves developing intercul tural communication abilities and using these to adequately manage intercultural situations and interactions. Recognition starts with knowing one's own culture and values and respecting others. Respect in itself is culturally determined. Uniting, bridging gaps, reconciliation and acknowledging require two-way processes.

> *How shall I talk of the sea to the frog*
> *If it has never left its pond?*
> *How shall I talk of the frost to the bird of the summerland*
> *If it has never left the land of its birth?*
> *How shall I talk of life with the sage*
> *If he is prisoner of his doctrine?*
>
> Chuang Tzu,
> 3rd century BC

Intercultural sensitivity can be increased by investigating the considerations made by Hofstede and by adopting an open and professional attitude towards the other culture. It is important to:

- Be aware of your role and position;
- Be aware of your own culture-specific background and how your culture defines your values and actions;
- Start a dialogue on intercultural aspects, similarities and differences;
- Find out how decision-making processes work; and, especially,
- Be yourself!

Advised communication techniques for water professionals:

- Many cultures are "oral" and "story-telling" cultures; use (the right!) metaphors and stories to convey your message.

- Use open (not suggestive) questions in order to clarify information.
- Be aware of body language, which may compensate for your lack of knowledge of a culture. Active listening is mandatory!
- Show respect! Be sincerely interested in the other person.
- Not all is culturally defined, but "human"-defined.
- Motivation is culturally based.

In practice this means that professionals must address at least these following issues (also taking gender differences into account!):

Water as a way of life:

- What is the history of the water source in this community/ region/ country? Local stories can tell you more...
- How did the community adapt itself to living with water under its specific circumstances and cope with changes in the water system in the past (centuries)? What traditional practices have worked well and what haven't? Can we use them?
- Questions about economic activities, gender, etc.

Water as a source of inspiration:

- What does water mean in the prevalent religion(s), spiritual beliefs and non-religious ideas, and do these set particular conditions?

Water as a source of power:

- Who is in control of the water and its management?
- What interests play a role?
- Who decides on important issues in this community? What are the roles of the others?
- (See also questions below.)

Water as a source of cooperation:

- Who owns the water source?
- When this new water (management) facility or system is installed, who will benefit? Who will be disadvantaged?
- What will be the consequences of it for the way this community functions? What will be advantages and disadvantages for the particular culture? What is important for the locals to keep of their culture? What are they willing to give up?
- How does water contribute to cooperation between communities or people within these communities?
- How are conflicts surrounding water issues normally solved in the community? What are accepted conflict strategies in this culture?
- Who is the person in the society who decides; how do decision-making processes work?
- What cultural aspects will be lost if this water system or facility is constructed or removed?

6.7 SUCCESS FACTORS FOR SUSTAINABLE INTERCULTURAL WATER MANAGEMENT

Realize that 'what works in one place doesn't necessarily work in another'

Implementing one approach world-wide doesn't work. New plans should fit into the local context. Recognize cultural factors and local interests, and factor them into your project.

Understand and respect local cultural values and beliefs

Appreciate the fact that cultural values and beliefs may differ from your own set of values and beliefs. Don't impose your own beliefs and values on others.

Build upon what has worked successfully in the past

Looking into the ways people used to adapt themselves to local circumstances offers some important lessons to help solve current and future water challenges in a sustainable way.

Consider the cultural impact of new water management plans

Make sure new water infrastructure or management systems have a positive cultural, socio-economic and environmental impact.

Involve local people

Use local experts and labour, and include local knowledge in your plans. Listen to them and respond to their ideas, concerns and needs. Remember that public participation is not the enemy of efficiency.

Think beyond barriers

People with different cultural backgrounds can shed different lights on problems, which can lead to innovative solutions. Therefore, use these differences as an inspiration to create new sustainable solutions.

Ensure a match between people having to work together

Select intercultural competent people. Create a positive and co-operative working atmosphere. If any misunderstandings arise, deal with them immediately and openly, thereby remembering that there is no right or wrong, but simply a difference in cultural viewpoints. Perform a stakeholder analysis.

Say what you do and do what you say

Regular, open and honest communication prevents delays caused by opposition and legal procedures. Be clear and specific in what you mean and what you plan to do, and don't make promises you can't keep or fail to follow up on.

See also Annex 1

6.8 CONCLUSION

In this book we have highlighted the role of culture in relation to water management and underlined the need for more cultural consciousness among water professionals. The water professional must take culture into account and investigate whether cultural values have an enabling or impeding influence in achieving sustainable water management goals.

The first step in this process is to understand what culture is and that it is more than artefacts, such as statues or folkloristic manifestations. Cultural values are important, as they are one of the factors that lie at the core of human behaviour. Neglecting these cultural factors means an omission in professional water management, especially because water use and water management are so culturally shaped.

Public participation and stakeholder management are indicators for success. If dialogues with stakeholders include the cultural factors at hand, sustainable water management will be achieved more easily.

Furthermore, awareness of culture may mitigate unknown incompetence, leading to the destruction of, or negative impact on, cultures (and thus people's identities).

Increasing one's intercultural sensitivity and intercultural communication skills will result in more effective water programmes. Last but not least, cultural diversity constitutes a wealth of knowledge, views and practices, that can be a source of inspiration, to create new sustainable solutions for water management.

Back to the future...

Corn Island Newspaper, 22 March 2020!

Sustainable approach stimulates economy, tourism and public health

The Caribbean Corn Islanders are proud of their international United Nations award for being the most sustainable island in the world. Following the cradle-to-cradle principle, the islanders use locally produced materials, and the transport of supplies from mainland Nicaragua has been reduced to a minimum. Everyone now has access to safe drinking water and sanitation facilities. Waste products are recycled into new products such as biogas and compost. The improvement of the drinking water and sanitation situation, supported by a foreign aid programme, provided the impulse for this integral and sustainable approach.

It's hard to imagine these days that only 12 years ago almost 70% of Corn Island's 9,000 inhabitants were not connected to the public water supply services and lacked adequate sanitation facilities. Waste was lying everywhere, which didn't make a pretty picture for the occasional tourist visiting the island, and it formed a serious health risk for the children. Domestic waste water regularly flowed into the drinking water wells, resulting in many health problems. Additionally, there were problems with coastal erosion. Many homes along the shore were threatened because the sea regularly flooded the shallow marshes, thereby increasing their salinity. At the time, these coastal marshes formed the most important source of drinking water.

With the financial support of the Dutch government, artificial reefs have been created that protect the coast and prevent further erosion. The sandy beaches have become larger, and new habitats for our rich marine life have been created. The good news spread fast and we now have many snorkelers and divers who visit Corn Island to explore the shallow and deeper reefs. But diving tourism is not the only attraction of our island. Since we received our award the eco-friendly hotels and guest houses are always fully booked by overseas guests. Everyone wants to come and see the eco-economic development with their own eyes and find out how we did it. From a forgotten tropical paradise in the Caribbean Sea, Corn Island has become a source of inspiration for a sustainable approach.

Waterschap De Dommel, The Netherlands
Louis Bijlmakers, 2008

Annexes

ANNEX 1 DO'S AND DON'TS FOR SUCCESSFUL INTERCULTURAL WATER MANAGEMENT.

Issues to consider	*Do's*	*Don'ts*
What works in one place doesn't necessarily work in another.	– Take cultural and local differences into account.	– Use one approach world-wide.
Adopt local traditions and practices into sustainable solutions.	– Try to build on what has successfully worked in the past.	– Consider traditional knowledge and practices as "backward".
Think global, act local.	– Involve local people in the planning process. – Consider the broader context and consequences of new plans. – Ensure the well-being of the local community.	– Forget that local issues need local input. – View your plan in isolation. – Forget to address the needs of local people.
Ensure a match between people having to work together, and think beyond barriers.	– Create a positive and co-operative working atmosphere. – Use cultural differences as an inspiration to create new sustainable solutions.	– Create an atmosphere of conflict. – Let cultural differences become a source of conflict that hinder the process.
Recognize cultural differences and local interests and factor them into your project.	– Find out what cultural factors (power distance, social relationships, knowledge level, etc.) determine the success of the project.	– Fail to ignore culturally-dependent enabling and counteracting forces.

&

Create local support for new plans.	– Involve local stakeholders in the decision-making process. – Visualize the situation to share conceptual understanding. – Value local people's suggestions and use them if feasible.	– Believe that public participation is the enemy of efficiency. – Think that you know best what is right for the people concerned. – Disregard suggestions of "lay people"
Understand and respect local cultural values and beliefs.	– Appreciate the fact that cultural values and beliefs may differ from your own set of values and beliefs.	– Impose your beliefs and values on others. – Assume you know what people think and want.
Listen to concerns and respond appropriately.	– Address the needs and concerns of local people seriously.	– Ignore or overrule people's needs and concerns.
Think ahead.	– Before starting a technical project make sure that the legal, financial and personnel responsibilities for long-term operation and maintenance are clear and covered. – Be pro-active.	– Trust that once realized, local people will use and maintain the system themselves. – Wait for problems to surface.
Use local experts.	– Involve local people in the work and create jobs for them.	– Try to do everything with your "own work force"
Regular, open and honest communication prevents delays caused by opposition and legal procedures.	– Say what you do and do what you say. – Make sure that your communication is line with the audience; use understandable language.	– Make promises you can't keep or fail to follow-up. – Fail to take language barriers into account.
Evaluate the project on a regular basis.	– Learn from your mistakes.	– Forget to evaluate the process, thereby not allowing for mid-course corrections.

ANNEX 2 LEWIS'S INTERCULTURAL COMMUNICATION MODEL

Another practical tool that will help build cultural bridges through understanding cultural differences is the one developed by Lewis. The Lewis Model of Culture is a most practical theoretical approach to classifying cultures and is quite useful for encouraging effective intercultural communication (see www.crossculture.com).

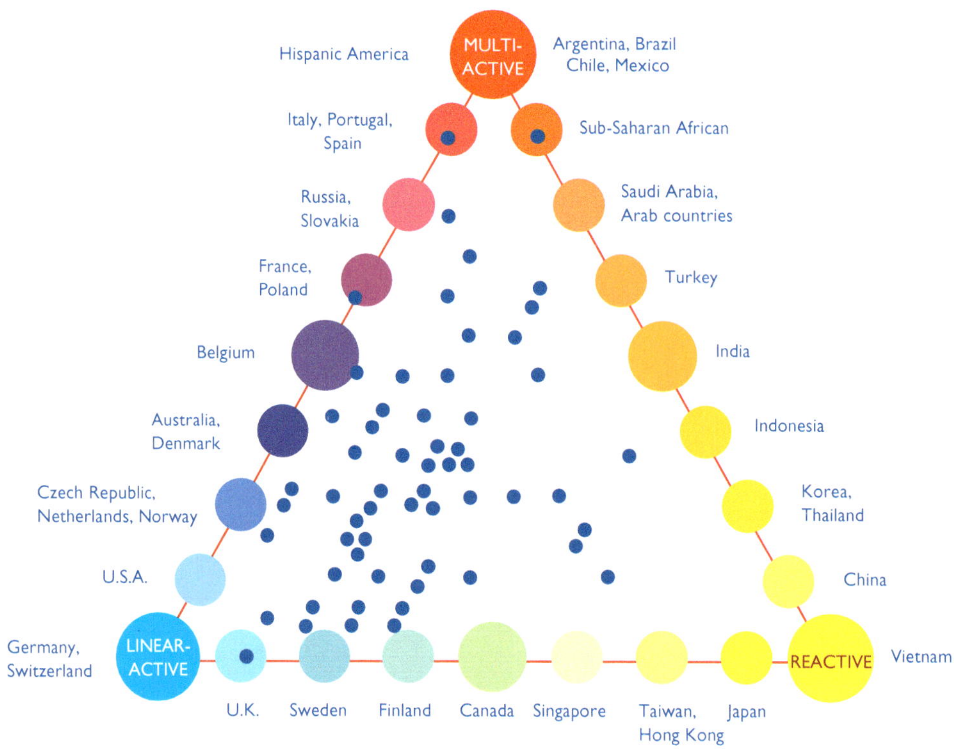

ANNEX 3 REFERENCES AND WEBSITES

1 Introduction to water and culture

[1]Third World Water Forum, 16–23 March 2003, Kyoto, Shiga and Osaka, Japan, Theme "*Water and Cultural Diversity*", Statement to the Ministerial Conference, 2003.

[2]Barr, A. and Serra, D. *Culture and Corruption*, GPRS-WPS-040, Global Poverty Research Group, University of Oxford, 2006.

[3]*Culture and the corporate priorities of the World Bank,* Report on Progress, 2003. and speech of James D. Wolfensohn, 1998 and *Paper on cultural and Sustainable Development,* World Bank, 1999.

[4]Llewellyn, D. – BA (Hons)/LLB (Hons), *BALAYI: Culture, Law and Colonialism* – Volume 7, 2005. Law School, University of Melbourne and Tehan, M.N – Associate Professor, Law School, University of Melbourne See also: *http://www.law.indiana.edu/students/centers/culture.shtml* on the relationship between Law, society and culture, 2005.

[5]Hofstede G., *Allemaal andersdenkenden, Omgaan met cultuurverschillen,* Uitgeverij Compact, Amsterdam, 1995.

[6]Glanz, J. and Miller, T.C. Official History Spotlights Iraq Rebuilding Blunders in *The New York Times* (December 13, 2008), web reference: *http://www.nytimes.com/2008/12/14/world/middleeast/14reconstruct.html?em*

[7]Rosenberg, M. *Polders and Dykes of the Netherlands*: The Reclamation of Land in the Netherlands, 2008.

2 Water: a source of life

[8]De Graaf, S. ed. *Water Stories*, Publication IRC International Water and Sanitation Centre, Delft, 2003.

[9]Ganeri, A. *Ancient civilizations series*, Compass Point Books; 1st edition, 2007.

[10]Roth, D., Boelens, R. and Zwarteveen, M. *Liquid Relations, Contested Water Rights and Legal Complexity*, Pub. Rutgers University Press, UK, 2005.

[11]IPPC, Climate Change, the Himalayan Mountains,in: *Sustainable mountain development,* Vol. 53, ICIMOD, 2007

[12]Photo web reference: *http://peopleandplanet.org/cms_graphics/img555_size2.jpg*

[13]Uijterlinde, R.W., Janssen, A.P.A.M. and Figuères, C.M., *Success Factors in self financing local water management*, Pub. Dutch Association of Water Boards, The Hague, 2003.

[14]Marchand, M. *Een Bengaal hoeft geen dijk* in: Newspaper '*De Volkskrant*', December 15, 2007.

[15]Water exposition, Dubai Museum, 2007.

[16]Shirazi, S.A.J. *Karez Irrigation in Balochistan*, web reference: *http://pakistaniat.com/2006/09/20/karez-balochistan-pakistan-irrigation/* September 20, 2006.

[17]Personal Communication Ing. W.van Dijk, water specialist member of the reconstruction team of Dutch Ministry of Defense in Uruzghan, Afghanistan, 2007.

[18]Skatssoon, J. *Aboriginal people built water tunnels*, web reference: *http://www.abc.net.au/science/news/ancient/AncientRepublish_1590192.htm*, ABC Science Online 15 March 2006.

[60]Alliance of Religions and Conservation (ARC) – Faiths and ecology – *Baha'i*; *http://www.arcworld.org/faiths.asp?pageID=2*. 2003.

[61]Dahl, Arthur. L. (1997) *The Baha'i perspective on water*, 2nd Klingenthal Symposium; WATER, Klingenthal, France, 26-30 November 1997; web reference: *http://www.bahai-library.com/conferences/water.html*, 1997.

[62]Redmond, C. *The water page* – Water in religion – *Baha'i*, web reference: *http://www.africanwater.org/religion.htm#Bahai*, 2000.

[63]*Unknown*, December 21, 2012 – are you ready? In: *India Daily, 2nd of Aug. 2006*, Web reference: *http://www.indiadaily.com/editorial/12485.asp*, 2006.

4 Water: a source of power

[64]Hansen, Roger D. *Water wheels*, web reference: *http://www.waterhistory.org/histories/waterwheels/*, 2004.

[65]Hansen, Roger D. *Water-related Infrastructure in Medieval London*, web reference: *http://www.waterhistory.org/histories/london/* (undated).

[66]PuddleandPond, *Water wheels*, web reference: *http://www.puddleandpond.com/water_wheels/water-wheel-history.htm*, 2007.

[67]Image: *http://energy.saving.nu/hydroenergy/images/howworks.jpg*

[68]Mohr, J. *Teatro del Agua: The Seawater Greenhouse "That Can Change the World"* web reference: *http://cleantechnica.com/2008/06/09/teatro-del-agua-the-seawater-greenhouse-that-can-change-the-world/*, 2008.

[69]McCully, P. Dams: What They Are and What They Do, from Chapter 1 of: *Silenced Rivers: The Ecology and Politics of Large Dams*, web reference: *http://www.internationalrivers.org/en/node/477*, (undated).

[70]Wikipedia, *Three Gorges Dam*, Web reference: *http://en.wikipedia.org/wiki/Three_Gorges_Dam*, 2008.

[71]Image: *http://www.export.gov.il/Eng/_Uploads/3575dum.jpg*

[72]Hvistendahl, M., China's Three Gorges Dam: An Environmental Catastrophe? In: *Scientific American*, March 25, 2008, Web reference: *http://www.sciam.com/article.cfm?id=chinas-three-gorges-dam-disaster&print=true*, 2008.

[73]International Rivers. People-Water-Life, *Frequently asked questions about dams*; web reference: *http://internationalrivers.org/*, 2008.

[74]Thaiportal, *Dammen verdrijven Birmezen*, web reference: *http://www.thaiportal.nl/index.php?option=com_content&view=article&id=10363:dammen-verdrijven-birmezen&catid=139&Itemid=100017*, May 22nd, 2007.

[75]International Rivers *New Hydropower Dam for Burma's Military Capital to Displace Thousands*, web reference: *http://www.internationalrivers.org/en/node/3105*, Full report and press release at *www.salweenwatch.org*, 2008.

[76]"Alles wat ik ooit kende, zal verdwijnen", Newspaper article in: *De Volkskrant*, PCM Publishers, 19th of May 2008.

[77]Unknown author *Frequently asked questions about dams*; at website of International Rivers, web reference: *http://internationalrivers.org/*2008.

[78]The World Commision on Dams, *A new frame-work for decision-making*, web reference: *http://www.dams.org/*

[79]Newspaper article "*China helpt aan waterkracht*", NRC Next 2 May 2008

[80]BBC News (April 21, 2008), *Africa plans biggest dam project*, web reference: *http://news.bbc.co.uk/2/hi/business/7358542.stm*

[81]*What is the Nieuwe Hollandse Waterlinie*: Web reference: *http://www.hollandsewaterlinie.nl/index.asp?id=19&lang=en*
[82]Westerman, F. (2007) Engineers of the Soul (*Ingenieurs van de ziel*), Amstel Uitgevers BV, 2007
[83]Wikipedia (2008), *Danube-Black-Sea canal*, web reference: *http://en.wikipedia.org/wiki/Danube-Black_Sea_Canal*
[84]Worster, D. Rivers of Empire: *Water, aridity and the growth of the American West*, New York, Oxford University Press, 1985.
[85]Hofstede, G. *Allemaal andersdenkenden, Omgaan met cultuurverschillen*, Uitgeverij Compact, Amsterdam, 1995.

5 Water: a source of cooperation or conflict?

[86]Bruns, B.R., Ringler, C. and Meinzen-Dick, R. *Water Rights Reform: Lessons for Institutional Design*, k, IFPRI, Washington, 2005.
[87]Donahue J. M. and Johnston, B.: Water, Culture and power, Island Press Washington, 1998.
[88]Greaves in Donahue J.M. and Johnston, B. *Water, Culture and power*, Island Press Washington, 1998.
[89]See: *http://www.who.int/water_sanitation_health/rightowater/en/*
[90]WHO, The right to water, World Health Organization. *Health and human rights publication series*; no. 3.ISBN 92 4, 2003, 159056 4 (NLM classification:WA 675), 2003.
[91]See also: *http://www.ipcri.org/watconf/papers/chatel.pdf*
[92]Wolf, A. Indigenous Approaches to Water Conflict Negotiations and Implications for International Waters, in: *International Negotiation* 5: 357–373, 2000.
[93]Donald, D.A. and Ruiters, G. eds., *The Age of commodity. Water privatization in Southern Africa*, Sterling: Earthscan, 2005.g, Va.: Earthscan, 2005.
[94]Hammoudi, A. Water rights and water distribution in the Dra Valley, in: *Property, Social Structure, and Law in the Modern Middle East, by Mayer, A.E.* SUNY Press, *State of New York Pr, 1985.*
[95]Muukkonen, S. *Water management in Cambodia-* Resources and Relations, University of Helsinki, Dep. Of Geography, January 2007.
[96]Kaboré, Daniel *Conflicts over land in the Niger river delta region of Mali: exploring the usefulness of SAM and CGE models to study participatory natural resource management in agricultural and pastoral systems,*Dissertation, University of Groningen, September 2008.
[97]Ludovic de L ys, H, Guinea, Mali, Liberia, Burkina Faso, Ghana and Cote d'Ivoire, UNOCHA, 2003.
[98]Chatikavanij, K., Patterson, F.W. and Lockett, W.G. The determination of power benefits from a multipurpose water project, *Engineering Journal of Canada*, Vol 53, no 12, p 21–27, Dec, 1970.
[99]Data from the Pacific Institute for Studies in Development, Environment, and Security database on Water and Conflict (Water Brief), 11-10-2008.
[100]Abu-Sitta, S. *Traces of poison*, article in Al-Ahram Weekly, Issue nr. 627, 5th of March 2003.
[101]Wolf, A., Kramer, A. and Carius, et al., Water can be a pathway to peace, not war, in:. *State of the World 2005 Global Security Brief #5*: on June 1, 2005

[102]Wolf, A. and Kramer, A. Carius, et al, Water Can Be a Pathway to Peace, Not War, in:. *State of the World 2005 Global Security Brief #5*: on June 1, 2005

[103]Jägerskog, A. The Jordan River Basin: explaining interstate water cooperation through regime theory, *Occasional Paper No 31, Water Issues Study Group (SOAS)*, University of London, 2001.

[104]Wikipedia, The Economic Community of West African States (ECOWAS) is a regional group of fifteen West African countries, founded on May 28, 1975 with the signing of the Treaty of Lagos, 2008.

[105]Kameri-Mbote, P. *Navigating Peace Policy Brief nr: Water, Conflict, and Cooperation: Lessons* From the Nile River Basin, for the Environmental Change and Security Program's Navigating Peace Initiative no 4, Woodrow Wilson International Center for Scholars, January 2007.

[106]Wolf, A., Kramer, A. Carius, et al., Water can be a pathway to peace, not war, in: *State of the World 2005 Global Security Brief #5*: on June 1, 2005.

[107]See also: *http://www.nattecanon.nl*

[108]Kameri-Mbote, P. *Navigating Peace Policy Brief nr: Water, Conflict, and Cooperation: Lessons* From the Nile River Basin, no 4, Woodrow Wilson International Center for Scholars, January 2007

[109]Myers, M. and Filner, B. *Conflict Resolution across Cultures: From Talking it out to Third Party Mediation* by 997, Amherst Educational Publishing. Amherst, MA ISBN 1-883998-19-0.

[110]Gleick, P.H. *Water Conflict Chronology*, Data from the Pacific Institute for Studies in Development, Environment, and Security database on Water and Conflict (Water Brief) 11/1020/08.

[111]Donahue J.M. and Johnston, B. *Water, Culture and power*, Island Press Washington, 1998.

[112]Nuiver, H. and Reijerkerk L.J. et al, *Verbinden met vertrouwen*, van Gorcum Publishers, 2007.

6 Water: a source of sustainability

[113]Keizer, H. Surinam Odo from: De *mooiste Surinaamse Mythen en Sagen*, Verba Publishers, 2004.

[114]Brundlandt definition, 1987 and World Commission on Environmental Development, 1987.

[115]*Success factors in self financing local water management*, NWB, 2003.

[116]Gupta, J., Glocal *Water governance; controversies and choices*, Delft, The Netherlands, 2007.

[117]Wijk-Sijbesma, C. van, *Gender in Water Resources Management, Water Supply and Sanitation*, IRC International Water and Sanitation Centre, Delft, The Netherlands, 1998.

[118]Sagnia, B. K. Framework for Cultural Impact Asessment, *International Network for Cultural Diversity* (INCD), 2004.

[119]Deming, W.E. *Management circle for continuous improvement* (1950's).

[120]*Culture Shock! Series* published by Time Media Private Ltd, see: www.expatriats.com